Piper Cub Tales

Frank B. Baker

First Printing 2005
Second Printing 2006
Third Printing 2012

ISBN 0-9788260-3-5

Eau Claire Printing Inc.
Eau Claire, Wisconsin

Acknowledgments

Many people have played a role in getting and keeping the Cub flying. My brother, John, whose initial investment made it all possible. Wally Watson, of the local EAA chapter, did all the critical welding that brought the fuselage back to its original configuration. The late Dan Spach, an A&I, did the initial inspections and licensing of the Cub. My across the street neighbor, Burr Fontaine, has been involved with the Cub since I first brought it home in a sorry state. His woodworking and machine shops have played a critical role in the Cub's life. My friends from the Madison Area Radio Control Society, Dick Buescher, Bill Vogelsang, Craig Lovell and the late Owen Kampen, have all been there to help me when additional hands were needed to accomplish some necessary task. In particular, they have hauled the Cub or its engine to and from the airport numerous times over the years. Dick Knutson, an aircraft restorer extra ordinaire, has been a constant source of information and advice about restoring and maintaining the Cub over the past 35 years.

I am deeply indebted to the farmers in whose fields the Cub made unannounced landings due to bad weather or a mechanical problem. In every case they were understanding about our plight and provided considerable help. In most cases, I was able to fly the Cub out of the field and head for home.

Over the course of the past decade, a number of people have contributed to the refinement of the many drafts of this book. My sister, Donna B. Lutz, reviewed many of the early chapters and made a wide range of insightful comments. My niece Robin Testa also reviewed the book and provided me with considerable thoughtful advice. Bernie Neuman has read the complete manuscript in its several different forms and contributed mightily. Bill Feeny, an artist and writer, has made valuable suggestions about the structure of the book's chapters and the flow of the content. My daughter, Ann, who was a participant in many Cub adventures, has played an active role in developing the book since its inception as a memento.

It was only through the contributions of these people that it was possible for me to fly the Cub over the many years and then put the experience to paper.

Forward

The Piper Cub was designed in the mid-nineteen thirties as a light weight, low powered, affordable, and easy to fly airplane. The resultant aircraft's characteristics: low landing speed; positive, but not sensitive controls; and minimal instrumentation made it an ideal training plane. During the late 1930's and into the late 1940's most persons who learned to fly did so in a Cub. As Cubs were used primarily for local flying they were often referred to as "puddle jumpers." Despite this attribution, cross-country flights in Cubs were common place.

I began flying my Piper Cub in the summer of 1969. The ink was hardly dry on my Private Pilot's License when my brother and I ferried the Cub cross-country to Denver, Colorado. Thus, began my focus on cross-country flying. On such flights, I fly low and slow, rely on only two instruments (compass and airspeed), navigate via map reading, and am subject to every whim of the weather. In addition, I have flown the Cub many thousands of miles cross-country without ever having communicated with a control tower, a Flight Service Station or (heaven forbid) Air Traffic Control. This is in sharp contrast to present day pilots who fly cross-country "in the system." In today's world of global positioning satellites, ARSA's, Mode C veils, and Air Traffic Control, such pilots fly high in the sky and have few degrees of freedom. While the Cub pilot is free of such man-made bonds and is able to savor the intimate view of the enfolding countryside only a few hundred feet below.

Over the course of some 35 years, I have made cross-country flights in the Cub to many parts of the United States. I have become intimately familiar with certain routes that I have flown multiple times. Farm houses, clumps of trees, railroad tracks and radio towers that I know along these routes become welcome sights as they mean the Cub is on course. On some of these routes, the weather is usually very nice. On other routes, I can count on running into low clouds, rain, and limited visibility. In the course of these flights, many adventures (some routine, some harrowing) have occurred.

My log books reveal that I have made many long distance cross-country flights alone in the Cub. However, I was struck by the fact that in the bulk of the chapters of this book, another person was with me on the flight. My companions were either a member of my family, a friend in the Cub, or I have been accompanied by a friend of mine flying his own airplane. It must be that the presence of a second person makes the trip more memorable.

For Christmas of 1993, I put some of these Cub adventures to paper, added photographs taken over the years and created a bound memento for each of my two children and one for myself. In the course of this project, it became apparent that flying a Piper Cub cross-country the way I and many others like me do represents a type of "seat of the pants" flying that is slowly disappearing from the aviation scene. When I retired in 1998, I set myself the task of transforming the memento into an actual book. It is my hope that the chapters of this book will convey to the reader some sense of flying a Piper Cub and why it is such a unique experience.

Contents

Chapter 1
Don't Buy a Piece of Junk!

In the summer of 1967. My wife, JoAnn, and I stopped in Rochester, Minnesota to visit Ed Schermerhorn, a long time friend of mine. As we chatted, Ed mentioned that he had a damaged Piper Cub, with no engine, stored in a barn at his parents farm. He was interested in selling the airplane for $200. That it was damaged did not register with me as being of great importance, but the price was attractive. The Piper Cub is a legendary airplane in aviation lore. It is a high wing airplane with tandem seating, a gross weight of 1220 lbs, and powered by a four-cylinder 65hp air-cooled engine. The airplane is covered with fabric and traditionally painted Piper Cub yellow with a black lightning stripe on the fuselage side. In the late 1930's and during the 1940's most people who learned to fly did so in a Piper Cub. Although the airplane is universally known as a Cub, its official designation is a Piper J-3.

JoAnn and I went back home to Madison, Wisconsin without looking at the airplane. Since one of the enduring legends of aviation is finding an old airplane in a barn, I began to wonder if I was missing an opportunity. After about a week of thinking about the airplane, I decided it was a "great opportunity." As a 40-year-old, university professor with a wife, two children, and a mortgage, I knew rebuilding an airplane would cost more than I could afford. I decided to involve my younger brother, John, who lived in Denver, Colorado. It was agreed that each of us would invest $1,000 and split any additional costs. I would rebuild the airplane and when finished we would be equal partners. His last words to me were "don't buy a piece of junk."

I telephoned Ed and told him I would buy the damaged Cub. I drove to Rochester, towing a flat-bed trailer, to pick up my new treasure. When I arrived, Ed and I went to the proverbial barn. The barn's roof was sagging and what little red paint remained had faded long ago. On the floor of the dimly lit barn sat a forlorn looking uncovered J-3 fuselage with a pair of covered wings leaning up against it. Other aircraft parts were strewn about on the ground. The best news was that the barn's roof did not leak and everything was dry. Ed and I started pulling out parts and loading them onto the trailer. There were many items that I didn't know of what use they would be, but we loaded all of them on the trailer or into the station wagon.

Retrieving the Cub from the Barn

On my way home, I stopped at the Lodi, Wisconsin airport to show my prize to Dick Knutson who had purchased a well worn J-3 the year before. He had done a very nice job of restoring it to better than new condition. He was to be a valuable source of information throughout my restoration project. Dick called an aircraft mechanic, who lived near by, to come and inspect my prize. In a few minutes, the fellow appeared. He looked at the loaded trailer for a while. I asked him if the airplane could be rebuilt. Being rather taciturn, he replied "its possible" and walked off to his car. This did not do a lot for my confidence as I was slowly beginning to realize that the airplane was badly damaged. Ed had told me that the Cub had been in a hangar in northern Wisconsin when the weight of a 3 foot snowfall caused the roof of the hangar to collapse. In the process, a steel roof beam had fallen diagonally across the airplane crushing it in the vertical dimension.

Once home, I laid all the items on the driveway to inspect them. Three of the four wood wing spars were broken near the root rib end and many wing ribs were smashed. In addition, considerable tubing in the top and bottom of the fuselage was badly bent. The landing gear was bent in several directions. To compound matters, the airplane had been owned by a succession of Experimental Aviation Association (EAA) types building their own airplanes. They had stripped anything removable from the airframe. Fortunately, Ed was an experienced hand and had latched onto an extra partial wing panel, some additional tail feathers, and other miscellaneous items in anticipation of rebuilding the Cub. As the extent of the damage was discerned, I began to realize that I faced a major rebuilding project. I called John and described what we owned. His reply was a testy "Frank, I told you not to buy a piece of junk."

Although I had been building model airplanes since the age of seven, I had few of the skills that would be needed to rebuild a badly damaged airplane. I bought a copy of the old Civil Aeronautics Board's bible for how to repair tube and fabric aircraft--Civil Aircraft Manual-18(CAM-18). Two other sources of information were crucial to my learning about restoring airplanes: the *Trade-A-Plane* newspaper in which various aircraft equipment companies advertize and the parts catalogs of those companies. I spent many hours reading these sources to learn the vocabulary of the field.

One of the many things that I did not anticipate was finding a place to work on the airplane. Our garage was not heated and winter was approaching. The only viable place was in my basement workshop. My model airplane flying buddy and ex-WW II B-24 pilot, Owen Kampen, and I carried the fuselage down my basement stairs. Then we managed to slip the 19foot long fuselage, through a side door, into my 21foot long workshop. To this day, I am not quite sure how we did it. I looked things over and decided what factory drawings were needed. A letter was sent to Piper Aircraft telling them what I wanted. In a few days, the necessary blueprints arrived in the mail and it was time to start rebuilding the Cub.

Knowing that numerous bent tubes would have to be removed, the whole airplane was jigged up with bolted together 2 X 4's. Once firmly jigged, the damaged tubing was cut out. Using the blueprints, I discovered that my jigs were preserving the damage. The dimensions were all off and it was necessary to loosen the bolts of the jig. When the first nut was loosened, I heard a loud bang followed by a few groans as the fuselage assumed more or less its original form. New measurements were taken and checked against the blueprints. Surprisingly, the fuselage had sprung back too within a few 1/8'ths of an inch of the correct dimensions. Several bowed lower longerons were straightened using my foot. Fortunately, the only undamaged tubing in the upper cabin of the airplane was the front inverted V that holds the front wing spar fittings. This provided a correct reference point for the reconstruction of the fuselage.

In two weeks, I had all the pieces of replacement steel tubing cut to length and tack welded in place. Although I recently had completed a welding course, I knew that I needed an experienced aircraft welder to finish the job. After many false leads and a several month delay, the local EAA chapter came through with a member who was a certified aircraft welder. In one night, he welded all the tubing and wing attach fittings in place. With the major welding done, I proceeded to weld the upper cabin structure (called the bird cage) together and install it on the fuselage. In the process I burned holes in two of the critical fuselage tubes and had to weld them shut. For years I worried about those places, but they never have become a problem.

With the fuselage in reasonably good shape, attention now turned to the wings. When the fabric was stripped from the wings, I discovered that several generations of field mice had called the wings home. I also learned that mouse urine cuts through aluminum like a hot knife through butter. About half the ribs had sections that were corroded through. Additional ribs were damaged by the falling roof. The wings were too big to take down the basement, so where to work on them became a pressing issue. Although it was January, I decided to do the work outside on the concrete slab for the new back porch. The sun reflecting off the back of the house and off the slab resulted in a comfortable place to work. The two complete wings and the partial one were disassembled. This entailed pulling out hundreds of small brass nails that held the ribs to the wood spars. In addition, the drag and anti-drag wires, the compression ribs, and numerous small metal fittings were removed. The damaged ribs were rebuilt by bending aluminum strips around the damaged sections and then riveting them in place. The end product of this labor was a stack of ribs, a small pile of wing spars, and a large number of other component parts. In the pile of wing spars there were usable front and rear spars for a left wing. Thus, only the two right wing spars needed to be repaired.

After several days of careful use of a hand plane, a 12 to 1 taper was achieved on the ends of both right wing spars and the replacement sections of spruce wood. Resorcinol glue was brushed onto the surfaces to be spliced and then the spar pieces were clamped together. The resultant glue lines were very thin and perfectly straight, indicating that a proper splice had been accomplished. Working in the garage during the spring and early summer of 1968, the wings were reassembled. With the completion of the wings, all the major assemblies for an airplane were at hand and not much of our $2000 had been spent.

A major milestone--the initial assembly of the complete aircraft--was now approaching. First, the fuselage had to come out of the basement. Owen and I snaked the fuselage out of the workshop. When the fuselage went down the basement, the cabin bird cage structure, had been smashed down. Now, it was at its full height again and the fuselage did not want to negotiate the stairway. After a lot of wiggling and pushing, the fuselage went up the stairs and out through a gap in the wall of the back porch. Then, Owen and I proceeded to assemble the airplane in the back yard. The wings fit perfectly into the fuselage spar fittings, which was a great relief. Unfortunately, the cabin birdcage was about 3/4" lower than the top of the wing even though it had been constructed according to the plans. The whole birdcage had to be cut off and remade to the proper height. This was the only major problem encountered in the rebuilding. Now, the basic airframe was complete, but it was not yet a Cub.

The need for an engine was looming large. One day, Dick Knutson told me about a Continental A-65 engine that he had seen advertized on a bulletin board at Milwaukee's Rainbow Airport. I telephoned the owner and the price, $300, seemed right. My thirteen-year-old son, Christopher, and I drove to Milwaukee to look at the engine. It had been overhauled by the aviation school at Janesville, Wisconsin that has a reputation for doing excellent work. In addition, the engine had been used only 75 hours since its major overhaul. I paid the man for it. As we loaded the engine in the station wagon, he asked if I needed any other parts. I casually said that I could use some control stick grips. He said I was welcome to the Cub parts he had at home. We drove a few blocks to his house where he proceeded to fill the back of the station wagon with every single item that could be removed from a Cub. I had found the pot of gold at the end of the J-3 rainbow. All the hundreds of small things the homebuilders had previously stripped from the Cub now were in my possession. Unfortunately, there were no wheels. The underlying story was that a thunderstorm had torn a Civil Air Patrol Cub off its tie downs and rolled it into a ball. The government had written the airplane off the books, so he and some cadets had stripped all the parts off the airplane.

About this same time, I tracked down a fellow who lived in the small town of La Valle some 40 miles north of Madison. He had been a fixed base operator(FBO) near Detroit, Michigan from the 1930's to the 1950's. When he retired, he brought five semi-truck loads of airplane parts to La Valle and stored them in two barns on his small farm. I bought a pair Cub wheels complete with brake assemblies, tires, and inner-tubes. The latter were stamped October 1941 and as of 1998 they were still on the Cub. He also poked around and found some original Cub wheel cover plates for me.

When I bought the plane, Ed gave me a bill of sale signed by a pair of brothers who had used it as a cropduster. In the late 1940's and early 1950's J-3's were widely used in this role. Since that time, the hulk had passed through several hands and I was not sure if the paper work was in order. With all the work already invested in the airframe and the dubious past ownerships, I started getting nervous about something going wrong. The appropriate papers were sent to the FAA aircraft registration section in Oklahoma City. After several months I still had not heard from the FAA. I contacted a former student of mine, who now taught at the University of Oklahoma, and told him of my problem. He happened to have a student in one of his classes who worked for the FAA. She discovered that my paperwork had been mislaid and put it back into the proper channels. Within a week, the registration number N88354 was ours. John and I now were the legal owners of the airplane. Over the span of ten years, 1937 to 1947, 19,802 J-3's (including military versions) were built. The serial number of our Cub was 15972. It was rolled out of the factory on February 21, 1946; the year 6,320 Cubs were manufactured.

Now, it was time to cover the fuselage with fabric. Dick Knutson's parents lived in a very large house in Lodi built in the 1890's. He had his Chiropractic office in the lower floor. Behind the main house, was a carriage house from the horse and buggy days. The hayloft had been converted into a workshop with an oil fired stove. For several months during the winter of 1968/69 I spent two nights a week and weekends working there. The cockpit interior was covered first using Ceconite(Dacron) fabric. Next, the floor boards, rudder pedals, brakes, front seat, and the control cables to the rudder and elevator were installed. Covering the exterior of the fuselage involves slipping a Ceconite envelope over it, gluing the edges and then shrinking the fabric with a hot household iron. It was just like doing a large model airplane. The tail feathers and landing gears also were covered and the fabric shrunk. Then, all the fabric was given two coats of brushed on nitrate dope.

The hayloft had essentially no ventilation and after a short time the odor of nitrate dope would get very strong. Dick would come outside between seeing patients and I would step out into the night air to visit with him. When the fresh air entered my lungs, I would become light headed and unstable on my legs. Having doped model airplanes for years, I thought nothing of the odor and proceeded to go back into the loft and complete the night's tasks.

My son, Christopher, had a friend whose father owned an automobile dealership. His service department had a state-of-the-art spray booth. He agreed to my spray painting the Cub there on a Saturday. I was very grateful for the use of the booth as it avoided several months of waiting for good painting weather. It was going to be the first time I spray painted a large object. Starting Saturday afternoon, three coats of silver and one of yellow butyrate dope were sprayed on the fuselage, tail feathers and landing gears. After each coat of silver, I lightly sanded the paint with a sheet of 320 grit sandpaper. The spraying and sanding process was finished by1:00 A.M. In the light of the next day, I could see unacceptable streaks of silver showing through the yellow. The next weekend, it was back into the spray booth. Only a few hours were needed to put two more coats of yellow on the fuselage, tail feathers, and the landing gears. This time they all came out perfectly -- beginners' luck.

By now it was late spring 1969 and warm enough to work in my garage. A Ceconite fabric envelope was slipped on each wing panel, glued in place, shrunk, and the required two coats of nitrate dope brushed on. A wooden stand was constructed for each end of a wing panel, so that a panel could be rotated about its long axis. The next task was to rib stitch the wings using waxed flat cord that is similar to wide dental floss. This is a tedious and time-consuming task. A long needle is pushed through the fabric. The wing in the rack is rotated and the needle and many yards of lacing cord are pulled through. Across the rib, the needle is pushed back through and the wing is rotated again. The needle and cord are pulled back through. Then a knot is tied. There are several hundred of these rib stitches on each wing. Looking back on this, I don't understand how I was able to do it alone. Then, 2" wide cotton tapes were glued over the top and bottom of the wing at each rib location. These tapes were brushed with clear butyrate dope and sanded until they were smooth. Three coats of silver and two of yellow butyrate dope were sprayed on the wings in a single day. This task was completed on Memorial Day weekend. There never was another weekend in the whole summer where the humidity was low enough to spray dope again.

Now that all the major components were covered and painted, it was time to complete the fuselage. The gas tank was installed using the front and rear sets of top and bottom straps. Long bolts passing through the straps and through four small tubular fittings hold the tank in place. I had the good sense to install the bolts such that heads of the bolts were welded to the top straps. When the boot cowl was attached, the gas tank strap bolt heads were hidden from view and inaccessible. However, the threaded part of the bolts would be accessible. A band saw was used to cut the clear plastic windshield to size. A nerve wracking and potentially expensive operation, however, it fit perfectly. I also installed a sheet of green colored plastic across the roof of the cabin in place of the traditional small skylight in the middle of the cockpit overhead. My sixteen-year-old daughter, Ann, and I hoisted the engine into place and bolted it to the engine mount. In the pile of miscellaneous items I had obtained from Ed, was a military Piper L-4 instrument panel. It was installed instead of the traditional curved J-3 panel. With it, the instruments were shock mounted and there was a bit more room between the back of the instruments and the fuel tank. Besides, it looked neat. Once these tasks were completed, the lightning strip and N88354 were painted on the sides of the fuselage using black model airplane dope. Then, the rudder, stabilizers, and elevators were installed. The final touch was placing genuine Piper Cub bear decals on the vertical fin.

Arrangements were made for a tie down spot at Morey Airport located just on the west edge of Madison. Owen and I tied the Cub's tail wheel to the tailgate of my station wagon and pulled the Cub to the airport on its own wheels. The wings were gingerly tied on the car's roof rack and taken, one at a time, to the airport. Owen, my neighbor Burr Fontaine, and I put the wings on and installed the wing struts. In mid-June, Dan Spach (an A&I) gave the airplane a final inspection. In the process, he found that I had left the cotter pins off all four bolts holding the wings to the fuselage. In the course of looking at the paper work, Dan also found that the airplane had been a cropduster and its air-worthiness certificate was for the experimental category. Even though he was an A&I, he could not change its category to that of a standard aircraft. To do this, it was necessary to get the FAA to inspect the airplane. After a delay of several weeks, a fellow arrived, did the inspection, signed the necessary papers, and gave me a temporary air-worthiness certificate. The appropriate papers were sent to

Oklahoma City. In mid-July, John and I had the permanent certificate and the Cub was completely legal.

A month after taking the Cub to the airport, the long anticipated day had arrived -- the Cub was to be test flown. Ken Poley (another member of our model airplane club) was an Air National Guard F-102 pilot. As a teenager he had flown J-3's from the Piper factory in Lock Haven, Pennsylvania to the people who had purchased them. He agreed to test fly the Cub. On Saturday, July 26, 1969; Ken, Owen, my family, and I went to the airport. Ken climbed aboard the Cub, I flipped the prop, the engine started, and he taxied to the runway. He took off, made several high and low passes as well as a few steep turns and landed. Then, he gave me a short ride in the Cub. The only problem encountered was that the engine would only turn up 1900 RPM instead of the nominal 2150 RPM. This was caused by the propeller having too high a pitch. A few years later it was also determined that the tachometer reads 120 rpm too low. The root of my propeller problem was that the fellow in LaValle said he had a Sensenich 72-42 prop for an A-65. He could not find it and sold me a Lewis prop, which he said was equivalent. Unfortunately, it was not. Later in the summer, I bought a Hartzel 72-40 propeller at the EAA Fly-In and installed it. This prop was excellent; good climb, 72 mph cruise and very smooth running.

At the beginning of the Cub restoration project, I was very unsure of myself. I really did not know what to do and lacked many crucial skills. As the project proceeded, The vocabulary of the field became understandable, the technical definitions were meaningful, and the blueprints made sense. Becoming increasingly comfortable in what I was doing; I could remove damaged tubing, do minor welding, and fabricate metal items. As I gained experience, I usually could sense when I was starting on a wrong path. Then, it was time to ask Dick Knutson how to do the task correctly. Covering and painting the Cub was the easiest part of the project. It was just like working on a huge model airplane. At the end of the two year process, I could look at the Cub with some pride of accomplishment. Despite John's admonition not to buy a piece of junk, he and I now were the proud owners of a beautiful Piper J-3 Cub.

Chapter 2
Learning to Fly the Cub

When I first brought the Cub home, I told JoAnn that I was only interested in rebuilding it. As the Cub neared completion, we both knew I could not resist wanting to fly it. The problem was that I did not have the necessary Private Pilot's license.

When I finished high school, my father gave me flying lessons as a graduation present. To this day, I can visualize flying with my instructor, Ray Hughes, who always wore an oil stained brown leather jacket. The cockpits of the Cubs and Aeronca Defenders used for training were very noisy and verbal communication between instructor and student was difficult. Consequently, Ray would indicate the direction of a turn by jabbing his lighted cigar in the desired direction. The cigar was his one and only training aid. After eight hours of dual instruction, including spins, I received my Student Pilot Permit.

Several of my high school pals also earned their Student Permits that summer. All of us lived and breathed aviation and were gung-ho on the air war in Europe. Renting airplanes at the same time, we would fly some distance from the airport, and then pretend to be fighter pilots. Some of these dogfights became quite "interesting" with two or three planes all trying to get on each other's tails. To escape an "attack", I once flew down the Minnesota River about 3 feet off the water. Much to my surprise, the branches of the big cottonwood trees along the bank closed in over the top of the Cub. After a few anxious moments, I found a gap between some trees and was able to pull up over the river bank and get out.

In the spring of 1945, I enlisted in the Navy V-5 pilot training program. I flew Stearman N2S biplanes at the Livermore, California Naval Air Station. My instructor asked me if I had flown before. I told him that I had logged about 20 hours in Cubs. We went up and he asked me to land the Stearman at a small grass auxiliary field. I made a nice pattern, came in over the barb-wire fence, rolled the tail wheel along the ground for a few feet, and dropped the main wheels on the ground with a solid clunk. He was surprised by my good imitation of a Navy style landing. He had me do three more and decided it was not a fluke. After that we didn't spend a lot of time on landings. I didn't tell him that I had been practicing such landings in a Cub for several months. Unfortunately, I made the mistake of being the first one in my class to solo whereupon they grounded me until the whole class moved onto pre-flight training at Ottumwa, Iowa. I logged 10 hours in the Stearman N2S and it was the highlight of my Naval Aviation career that was curtailed by the end of the war.

Despite my previous eleven months of military service, my draft board still had me classified as 1A, which is the first group to be called up. The Korean war started a few months after I graduated from college and I received my draft notice. Through a series of fortuitous bureaucratic maneuvers, I was able to be accepted into the U.S. Air Force navigator program. My class was 51-26 and training began in March of 1951at Ellington AFB, Houston, Texas. The training missions were flown in C-47's with two instructors and ten cadets in each plane. When the class graduated in September, I was commissioned a 2'nd Lieutenant. The whole class then went to combat crew training at Langley field, Virginia where we flew in Douglas B-26's. For reasons never made clear, I was part of a group that ferried six newly remanufactured B-26's from McClellan AFB in Sacramento, California to Japan.

In mid January, 1952; I arrived at K-14 AFB (Kimpo) near Seoul, South Korea and was assigned to the 6166'th Weather Reconnaissance Flight of the 12'th Tactical Reconnaissance Squadron. I flew a mix of day and night weather reconnaissance missions over North Korea and the China Sea. I also flew a number of night photo reconnaissance missions. By late May 1952, I had flown my 50 missions. My replacement was one of my navigation instructors from Ellington AFB. I had the pleasure of giving him his first check ride over North Korea at night along the Yalu River. The remainder of my tour of duty was spent at Ellington AFB as an instructor. By the time I was discharged from the Air Force, in August of 1953, I had accumulated 500 flight hours of which 176 were combat hours.

In May of 1969, inspection of my old student pilot logbook revealed that my last entry was dated June 2, 1948 when I had flown a J-3 at Annapolis, Maryland. Although I had accumulated 37 hours of flying time by then, I would have to start over again to obtain a Private Pilot's license. To this end, I started taking flying lessons at Morey airport. My instructor asked me if I had flown before and I told him that I had been a navigator. His reply was "that does not count." Clearly, I was back at square one. Despite this, he let me solo a Cessna 150 after only five hours and twenty minutes of dual instruction. I was never particularly happy flying the C-150 as it had a control wheel rather than a control stick. All my previous flying had been in "stick and rudder" aircraft, which I much preferred.

It wasn't until July 29,1969, after only 30 minutes of dual instruction in the Cub, that I went solo in it. In retrospect, soloing the Cub after only 30 minutes of dual instruction is quite unusual. I suspect that having flown a Cub some 20 years ago had ingrained the basics of flying it. Now that I could fly the Cub solo, the immediate task was to build up enough flying time in the Cub to meet the requirements of the Private Pilot's license.

Like most of the light planes developed during the 1930's the Cub's engine did not have an electrical system or a starter. Thus, it was necessary to start the engine manually. The procedure was: set the throttle to idle, pull the prop through five or six revolutions to introduce gasoline into the cylinders, switch the magnetos to the on position, and give the prop a quick downward pull. If all went well the engine would spring to life. To accomplish this safely, the process was accomplished while standing alongside the open door on the right side of the fuselage and behind the prop. While my left hand had a firm grip on the door frame. My right hand was free to pull the prop through. Cub pilots facetiously refer to this procedure as using an "Armstrong Starter".

After the Cub's engine is started and it is running at idle RPM, rather loud rattling sounds can be heard. When I first noticed these sounds, I was quite concerned that there was a problem in the engine. I talked to a couple of other Cub owners about the noise and they all agreed it was just "Continental clatter" that is a characteristic of the engine.

When flying the Cub solo, I sit in the rear seat and am well under the wing. This gives the impression that the I would have a limited field of vision. Actually, there is an exceptionally good panoramic view in more than an 180 degree arc. The right and left side windows are quite long and provide an excellent view out under the wings. Looking down out of the side windows, there is an good view of the ground except for what is directly under the fuselage. In a steep turn, the skylight in the cockpit roof provides some visibility in the direction the Cub is turning. Like all high wing airplanes, the visibility to the rear is quite limited and clearing turns are necessary to check on what is behind the plane.

A pair of cylinders on the A-65 protrude from each side of the cowl. On top of each pair is a black painted air scoop that is called an "eyebrow." When making a turn, the high side eyebrow is positioned on the horizon. If held there properly, no altitude will be lost while making the turn. It takes a while to learn the proper position of the eyebrow on the horizon as a function of the steepness of the turn, but it quickly becomes an automatic procedure.

A major focus was on becoming skilled at taking off and landing the Cub. The former is the easier of the two tasks. When flying the Cub solo, I sit in the back seat. On the ground, the nose of the Cub blocks my view of the runway straight ahead and only about 50 yards of the sides of the runway can be seen. Once the Cub is aligned with the runway, the throttle is advanced smoothly, and the Cub slowly builds up speed. At about 25 mph, a little forward pressure on the stick will raise the tail and the Cub is in a level position. Now, rest of the runway can be seen. At about 40 mph, a bit of back pressure on the stick will cause the Cub to break free of the ground.

Landings took a quite a bit of practice before I could executed them consistently. Full stall landings are reasonably easy as long as the timing of them is correct. The trick is to bleed off the airspeed a few feet off the ground and then gently pull the stick all the way back to stall the Cub at the same moment as the wheels touch the ground. Wheel landings require the Cub to be put in a crabbing position with a low upwind wing. The opposite rudder is used to maintain the heading along the runway. Touch down is made on the upwind wheel at a much faster rate of speed than for a full stall landing. As the speed dissipates, the other wheel touches, and eventually the tail wheel makes contact with the ground.

For either type of landing, once the wheels are on the ground, maintaining directional control becomes critical. The Cub's rudder is very effective and the tailwheel is connected to it. Since pushing on either of the rudder pedals has an immediate effect on where the nose of the Cub is pointed, the pedals must be used judiciously. If the nose swings to the left, a bit of right rudder is fed in. As soon as the nose starts swinging to the right, the pressure on the right pedal is eased, and the left pedal is used in a similar fashion. The result is that I do a tap dance on the rudder pedals. An observer watching a Cub land, sees the rudder quickly swinging from side to side. Given the least amount of inattentiveness, the airplane acquires a mind of its own. Many a neophyte Cub pilot has ended up ground-looping because of too early or too late use of the rudder pedals. There is a saying among Cub pilots " you are not through flying a Cub until it is in the hangar."

At this point in time, my flying was done in a rather tenuous manner. I was concerned as to what restoration mistakes that I might have made and what their consequences would be. As I built up time in the Cub, I became increasingly confident in the integrity of the airplane. However, I rarely made turns with a bank angle of more than 15 degrees and never with any discernable amount of G force. However, I did make it practice to fly in all kinds of weather rather than just on nice days. When the wind was blowing 15-20 mph, I would practice wheel landings in a significant crosswind.

In the summer evenings, one of the pleasant aspects of flying is that the sun does not set until nearly 9:00 P.M.. The air is usually cooler, the wind has died down, and the low sun gives the countryside a golden glow. The white farm houses stand out against the green fields as if a spotlight has been shined on each house. Many evenings, I would see deer walking along the edges of the woods as they came out to feed.

Part of the requirements for a Private Pilot's License is to have flown a short and a long cross-country flight alone. My short solo cross-country, flown in a Cessna 150, was from Morey Airport to Dubuque, Iowa, a flight of 79 miles. I flew at a cruising altitude of 2,500 feet. It was a clear day, so from this altitude I could see at least 25 miles in all directions and did not need to use any of the radio equipment. The major landmarks, such as the Wisconsin River and the numerous rural towns, were used as aids to navigation. Since the weather was nice, it was just a matter of pointing the airplane in the right direction until I arrived at my destination. Flying to Dubuque took 40 minutes. I landed, had my log book signed, took off, and headed back towards Madison. The return trip took an hour and ten minutes.

Three days after the Dubuque trip, I did the planning for a long cross-country flight, in the Cub, to Mankato, Minnesota where I could stay overnight with my parents. This flight involved a 102 mile segment to La Crosse, Wisconsin, refueling the Cub there, and then a 136-mile segment to Mankato. After reviewing my plans, my instructor signed me off for the trip. A hidden agenda was to stop at Rochester, Minnesota to show the restored Cub to Ed Schermerhorn.

When I was an Air Force Navigator, most of my navigating was accomplished via good old-fashioned map reading and computing estimated time of arrival (ETA) to the next checkpoint. Thus, I was confident that navigating in the Cub would simply be a repeat of those days. The course line was laid out in red on the Chicago and Omaha sectional maps. Every ten miles along the course line was marked off and the compass headings were written on the maps. Since I was going to fly over unfamiliar territory, I studied the maps for several days. Finally, I was familiar with the major checkpoints along the course line.

As the Midwestern weather can be very unpredictable, I checked weather for the next couple of days with the Lone Rock, Wisconsin Flight Service Station (FSS). It looked good, so at 8:00 A.M. on September 12,1969, I took off and for the first time headed the Cub over the horizon. It was both a thrill and a bit disconcerting as it was a flight into the unknown.

Once the Cub was off the ground, a climb was maintained until my cruising altitude of 2,500feet was reached and a northwest heading of 300 degrees was established. I flew past the west end of the Baraboo Bluffs that are a wooded 25 mile long, 500 foot tall glacial moraine. Every ten minutes the wind vector, ground speed, and the estimated time of arrival (ETA) to the next checkpoint were computed. Only two of the five instruments on the Cub's panel (airspeed, and compass) are involved in navigating--nothing fancy here.

The unglaciated country between the Wisconsin River and LaCrosse is a seemingly endless series of meandering valleys. At their north ends they are deep and narrow but become broader as they approach the Wisconsin River. All along both sides of a given valley there are numerous side valleys that join it. This results in ridges between the side valleys. All the valley sides and top edges of the ridges are densely wooded. There are small dairy farms both on the tops of the ridges and down in the valleys. In the bottom of some of the side valleys are creeks that join a larger creek in the main valley. This creek then flows south to the Wisconsin River. Usually there is a country road that follows the main creek. By this time of year, the tree leaves were beginning to turn color, the mature corn fields were light tan, and the alfalfa fields were still a lush green.

As I began flying over this unglaciated area, a ground speed of 40 mph was computed even though the indicated airspeed was 72 mph. After 55 minutes of plugging along at this ground speed, I computed an ETA to LaCrosse that indicated the city was over an hour away. I thought there was not enough fuel in the Cub's tank to fly the rest of the way to LaCrosse.

There were no airports on my line of flight between Morey's and LaCrosse, but the Sextonville airport was only a few miles off course to the south. I quickly drew a line on my map from my present position to the airport and took up the new compass heading. In about ten minutes, I could see the airport down in a valley and landed there.

The airport was managed by a FBO named Al. In the 1950's he owned a WW I Thomas-Morse Scout Biplane that he sold to the Talmantz museum in California. While he put gasoline in the Cub's tank, I told him about my ground speed problems. He asked:

"How high are you flying?"

I replied: " 2,500feet."

"That's too high for a Cub. If you can't see, the individual leaves on the trees,
you are too high."

He recommended that I fly just above the tree tops to take advantage of the attenuation of the wind due to objects on the ground.

After taking off, I flew north to pick up the original course line. After getting on course again and setting up the proper wind correction angle, I followed Al's advice. The Cub's ground speed jumped up to 55 mph, which was better, but navigation became a real problem.

Flying at this low altitude, there is a circle roughly two miles in diameter in which information on the map can be matched with the countryside. I quickly discovered that it was necessary to navigate by using the contours of the ridges, bends in the creeks, and occasional villages tucked in the valleys as checkpoints. The partially folded sectional map was perched on my thigh. An aviation plotter was used to measure distances on the map. A WW II war surplus E6B , a form of a circular slide rule, was used to compute the ground speed from the time and distance between check points. Of course, the map would not stay still on my leg and I only had one hand free to manipulate the plotter and E6B. This situation was further complicated by the Cub being tossed around by the wind bouncing off the faces of the ridges.

Once back on course, the first checkpoint was a ridge formed by the west side of a valley. When a ridge about a mile ahead was visible, I decided that it was the one that was on the course line. The ridge actually was a bit off the course line. Consequently, I started having difficulty matching what I was seeing to the anticipated contour lines on the map. Then, the ridge passed below and behind the Cub. Now, I did not know if what I had seen was north or south of my course line. A search process was begun in which I tried to match what I was flying over with where I thought I was on the map. Failing this, the west bank of the next valley was targeted and the process repeated. After about two or three of such incidents in a row, confidence in my navigating skills was fading rapidly. Circling around to get a second look never entered my mind.

About three-fourths of the way to LaCrosse, the Kickapoo River flows through a large valley. There is an end of a ridge just north of the village of La Farge that is good checkpoint. This ridge appeared just about the time I thought the Cub was way off the course. In reality, I was only a half mile or so off course. Having found a rock solid checkpoint renewed my confidence. To do a better job of staying on the course line, a slight adjustment in the compass heading was made. It was becoming painfully clear to me that navigating a B-26 at 10,000feet was a lot easier than navigating a Cub at 100feet above the trees. From the B-26, on high, I could see at least 25 miles and could use major terrain features as check points. At that altitude the air was very smooth, the B-26 was very stable, and my map of all of Korea was on my clip board. However, just as in the Cub, I relied on the compass and airspeed as well as my trusty E6B.

Once past La Farge, navigation became easier. A few miles past the hamlet of Newry, the tall ski jump at the village of Westby was silhouetted against the sky. At this point, the valleys became deeper and more heavily wooded. A few minutes past the ski jump, I could see the cluster of buildings at Newburg Corners and the light-colored gravel road in front of them. In a minute or so, the broad valley of the LaCrosse River could be seen to the north. This river flows from the northeast and joins the Mississippi River near the LaCrosse airport A few miles beyond Newburg Corners is the hamlet of St Joseph that is located on top of the south bluffs of the LaCrosse River valley. There is a white water tower and a large tan brick convent building on the north edge of the hamlet. Even when flying down low these can be seen from quite a distance. As I flew over the convent, nuns, in their black garb, could be seen tending their gardens.

The LaCrosse airport, located on a large island in the Mississippi River, is one of the most beautiful I have ever seen. Its white concrete runways contrast vividly with the green of the grass and the blue of the water. To the north and south, the river flows through a several mile wide valley. East and west of the river there are very steep, 300 foot high, heavily wooded bluffs.

The tall bluff on the east side of the Mississippi River valley has a stone quarry, on its top, that is right on the course line to the LaCrosse airport. On final approach I had to lose a lot of altitude in order to land on runway 31 that was far below. I started from above the quarry, passed over a branch of the Mississippi, and then over a short stretch of grass. A high rate of descent was maintained until it was time to flare. When my senses told me that I should flare, I pulled back on the stick. The Cub stalled about two feet too high and hit the ground with a resounding clunk. After rolling out, I taxied to the Viking fixed base operation.

After refueling the Cub, I taxied to runway 31 and took off to the northwest. It was a cool day, so that only a few S turns were needed before the Cub was higher than the tops of the bluffs. Once clear of the three or four radio towers perched on the bluffs, the white cone on the roof of the Nodine VOR building could be seen. Pilots in more modern airplanes have radio equipment that allows them to take a heading to Nodine based on electronic indicators on their instrument panels. I simply used the building as a visual reference and flew to the same point as they do, but at a much lower altitude.

After Nodine, there were three Minnesota villages that were excellent checkpoints despite their not being exactly on the course line. The Cub was flying over corn and hay fields and an occasional turkey farm could be seen. Low flying airplanes are the scourge of turkey farms. If turkeys see an airplane, they panic and run to the corner of a building where large numbers of them die from being crushed. Whenever I saw such a farm, I gave it a wide berth and always watched to see if the turkeys started to run. Fortunately, the Cub and I did not start a turkey stampede at any of these farms.

Because I wanted to show the restored Cub to Ed Schermerhorn, I had made arrangements to meet him at a private airstrip east of Rochester. About where I thought the airport should be, I started looking, but could only see green fields. I flew back and forth for a few minutes. Finally, a wind sock on a hangar was seen, so I proceeded to land on the grass runway. I showed the Cub to Ed and his friend. They were quite impressed at the final product. Ed knew what I had started with and admitted that he had given up on the plane. He was happy to have sold it to me. Ed and I went for a short ride in the Cub. He did some steep turns that gave me pause as I knew about those two wing spars that I had spliced

After taking off from the grass strip, I headed the Cub in the general direction of the Mayo Clinic in downtown Rochester. The two Clinic buildings are in the middle of the downtown area. The oldest is the Plummer building that is a tan colored tower about 13 stories tall. The newest building is much larger, taller and faced with white marble. This building is rather dramatic in appearance and can be seen from miles away even at the low altitude that I fly the Cub.

Once west of the Clinic, I flew the Cub a bit south of my plotted course line to follow Highway 14 to Mankato. Approaching Dodge Center, I was flying at about 200feet above ground level(AGL) and just off the north end of their airport's runway. My attention was in the cockpit while I computed a ground speed. Suddenly, something appeared in my peripheral vision. I looked up just in time to see the red light on the top of a radio tower pass by at what seemed to be about 3 feet below the left wing tip. It happened so quickly I could not react in time to avoid it. The red light was so close that I could see the ridges in the glass and the bolts that held it onto its base. My flying career almost ended on my first solo cross-country flight in the Cub. What a bummer that would have been. After calming down a bit, the map was checked very closely. The tower was not on the map. From then on, radio tower symbols, within 5 miles of the course line, were circled in red on the map, just like I did known anti-aircraft batteries in Korea. They were equally as deadly.

The rest of the way to Mankato, I followed the highway. Every seven to eight miles the Cub flew over a town that I knew well. Finally, the Minnesota River valley came into view. The city of Mankato is strung out along the south side of the river and up and over that side of the valley. Right in the middle of town and alongside the river is a large flour mill with its tall concrete storage bins. I flew a mile or so past the airport, circled over my parents house, went back to the airport, and landed. The whole flight, including a short hop with Ed took four hours and 54 minutes of flying time. I went into the airport office and the man at the desk signed my log book. This was done to prove that I had really gone to Mankato. I stayed over night in Mankato with my parents, who were suitably impressed by the newly restored airplane that John and I owned.

On the return trip, the leg from Mankato to La Crosse was uneventful. The leg from La Crosse to Madison, over the unglaciated area, still was tenuous. Now, I was looking at the east sides of the same valleys and it was a totally new view of the world. As a result, I worked very hard to keep the Cub exactly on course. Only a slight distance off the course line would result in a repeat of the problems I had on the way to LaCrosse. Going home took four hours two minutes of flying time. The return trip was 238 miles in length and flown at an average ground speed of 57 mph. One thing for sure- -the Cub is not very fast.

A few days after I came home, I flew to the Lone Rock Flight Service Station (FSS) and took my Private Pilot written exam, which I passed. Finally, on the 7'th of October 1969, I took my flight test, in a C-150, and received my Private Pilot's license. The event occurred some forty years after my mother handed me two wood dowel clothes' pins, one serving as the fuselage and the other as the wing, and said "see airplane." Now, I was a licensed Private Pilot and only time would tell if I would become a "Cub Pilot."

Chapter 3
Sharing the Cub

The First Ferry Flight: Madison to Denver

John and I had agreed that the Cub would be kept in Wisconsin from Memorial Day until the fall of the year. He would keep it the rest of the year. As the Denver, Colorado area is known for occasional very high winds, a stipulation was that the Cub be in a hangar there. Our sharing agreement necessitated ferrying the Cub from Madison to Denver and return each year. John had earned his Private Pilot's license flying a Cub in the late 1950's. By the present time, he had flown about 1,000 hours and was an experienced cross-country pilot. After some discussion, we decided that the first Madison-Denver ferry flight would begin in the second week of October 1969.

To prepare for the trip, I bought the Chicago, Omaha, Cheyenne and Denver sectional maps. The first leg of the planned flight was from Morey Airport to LaCrosse. The second leg was from there to Mankato, so John and I could visit our parents. The route then was from Mankato through southern Minnesota, northwestern Iowa, central Nebraska and northeastern Colorado to Denver—a distance of 929 miles. We called this the "southern route."

Southern Route from Madison to Denver

The Cub's gas tank holds only 12 gallons. With two people onboard, the engine burns about 4.1-4.3 gallons an hour. The gas gauge on the Cub is a cork float with a 3/32" diameter rod attached to it. The rod sticks up vertically through the gas cap located just in front of the windshield. The top of the rod was bent over, by the original owner of the Cub, so that after two hours of flying time, the horizontal part of the wire rests on the top of the gas cap. This is a visual warning that the Cub is low on fuel. In a pinch, another 20 minutes can be flown, but an unknown quantity of fuel remains. Thus, a prudent distance between gas stops on a cross-country flight is about 100 miles.

Course lines between airports about that distance apart were drawn on the maps. Then, every ten miles were marked on the course lines. The compass headings for each leg also were written on the maps. I studied the maps for several hours each night until I was familiar with the major features of the planned route.

In order to land at airports with a control tower, I had installed a ten channel Regency hand-held radio on a bracket attached to the back of the front seat. To use the radio, the person in the back seat leans forward and listens to or talks into the radio's combination speaker/microphone. Since there was no intercom system in the Cub, verbal communication between the two occupants involves a lot of yelling.

John arrived in Madison, the day I received my Private Pilot's license. We departed from Morey Airport at dawn on Wednesday, October 8, 1969. It was cold and dreary day with a high grey cloud layer. The lower right hand door of the Cub does not fit very well. There are several openings around the front of the door and daylight can be seen along the rear diagonal of the door. Despite having weatherstripped these openings, a breeze comes through when flying in cold weather. As a result, the temperature in the Cub's cockpit is the same as the outside air. Because of this, down ski jackets, wool socks and hiking boots were the uniform of the day.

John sat in the backseat, held the map on his thigh, and was the navigator. My piloting duties in the front seat were to hold the airspeed, altitude, and compass headings John gave me. After crossing the Wisconsin River, the Cub entered the un-glaciated area. The fall tree color here was just past peak, but in the dim light it was not very colorful.

This was only my third time on the 102 mile Madison to LaCrosse segment. At our altitude of 200feet AGL, each new valley looked just like the last one. Having John navigate, allowed me to spend more time getting acquainted with the terrain. Two hamlets were directly on the line of flight. The first was Hub City in the bottom of a wide valley. The second, Yuba, was further northwest in the same valley, but where the valley was quite narrow. This latter hamlet, composed of only a dozen or so houses and a few stores, is tucked on the valley floor around the southwest point of a ridge. On our line of flight it could be seen easily.

In the valley of the Kickapoo River, the town of LaFarge was a welcome checkpoint that appeared right on schedule. John said the map showed a settlement a few miles ahead, so I started scanning the area. The settlement turned out to be three small houses perched on the side of a ridge. A few miles further, the Cub passed over the reservoir south of Newry assuring me that we were on course. In a few minutes, the ski jump at Westby came into sight.

At the St. Joseph's convent, I deviated slightly to the north to get above the south side of the LaCrosse River valley. This position gave me a better view of the LaCrosse airport and any possible airplane traffic. There is a FSS, but no control tower at LaCrosse. All John had to do was announce our presence on the proper radio frequency. Then, I proceeded to land the Cub on the very long concrete runway. This was the first time that I had landed the Cub from the front seat and the visual picture was different from that of the back seat. Now, the wide concrete runway seemed to reach up to the horizon and it filled my field of vision. My estimate of how high the Cub was above the runway was distorted and I made a bouncy landing.

At LaCrosse, the Cub was refueled. Then, John and I switched seats and duties for the 136 mile leg to Mankato, Minnesota. After takeoff, John climbed the Cub towards the west river bluffs, but with two people onboard gaining altitude was a slow process. Finally, the Cub was high enough to clear the bluff in front of us. Then, we worked our way around the several radio towers perched on its top.

Initially, Interstate 90 was next to the course line, but it soon veered off to the southwest. Navigation checkpoints were provided by a few radio towers and a number of villages. One of these was St. Charles that is located behind a hill. Coming from the east, all that can be seen of the town is the top of its silver water tower barely protruding above the trees.

Since there were both a navigator and a pilot in the Cub, the straight line course from Rochester to Mankato was followed rather than Highway 14. This course is north of the highway and eventually goes directly over the town of Waseca. Approaching it, John put the Cub in a climb to get well above the town and once past it he put the Cub back down to just above the tree tops. Reaching Mankato, we flew over our parent's house before we landed at the airport. The only available tie-down was on the asphalt taxiway between two rows of hangars.

When we awoke Thursday morning, the wind was blowing at 35-40 mph--there would be no flying this day. John and I drove to the airport to check on the Cub. As we came around the corner of the hangar row, there was the Cub flying on the tie-down ropes with all three wheels off the ground. We hauled the Cub back down to the ground and tightened the ropes so that it could not lift off again. During the night, the wind subsided.

Friday morning, John and I were up at the crack of dawn to continue the trip with myself as pilot and John as navigator. The first gas stop was to be at Spencer, Iowa. When a pilot says that he is going to a town, he actually means that he is going to land at the town's airport. Often the airport is several miles from the town. Except to stay over night, the town itself is rarely visited. The 90 mile flight to Spencer is initially over flat farm land that transitions to rolling countryside in northern Iowa.

At 60 miles from Mankato, there is a long east west rise capped by three small towns that are about five miles apart. The task is to decide which one is on the course line and can be used as an aiming point. Landmarks seen at a distance often appear to be on course, but are not and they will lead you astray. John warned me about the problem, but I picked the wrong town. When the Cub reached the town, we were 5 miles off course. A slight correction to the compass heading put us back on track to Spencer. The late Shelby Hagberg had lived in Spencer. Over the years, he had collected a number of 1920's Curtiss Robin airplanes and a vast amount of parts for them. Many of the Curtiss Robins in existence today owe their survival to Shelby.

The Curtiss Robin has always been my favorite antique airplane. In the late 1930's there were several of them at the Mankato airport. I was the typical "airport kid"and spent quite a bit of time hanging around the airport. I even got a flight "around the patch" in a Robin as a reward for helping wash several airplanes. A ride that I have never forgotten. In the mid 1990's I tried to buy a "basket case" Robin that needed to be restored, but the negotiations fell through.

From Spencer to Sioux City, Iowa the countryside is gently rolling hills covered with golden tan colored fields of matured corn and soybeans. Looking down, the Cub's shadow could be seen flowing up and down across the fields and hills as if it were racing us to our next destination.

The airport at Sioux City, is on the Missouri river flats and has a control tower. Its 9,000foot long concrete runway was quite intimidating to this grass field pilot. John called the tower on our little radio. The tower operator would say something, I would yell at John "what did he say?" John would lean forward and yell directions in my ear. After about four such interactions I had the Cub on final approach and I made my second landing from the front seat.

The next leg was from Sioux City to Columbus, Nebraska---a distance of 82 miles. Once away from the Missouri River valley, the land is rather flat. Initially, there were plenty of navigation checkpoints, but as the Cub approached Columbus they started to get further apart. After refueling at Columbus, we took off with John in the front seat and headed toward Grand Island, Nebraska only 60 miles away.

A dual lane railroad track runs straight from Columbus to Grand Island. All John had to do was follow it. Coming into Grand Island, the sun was starting to disappear behind a low cloud deck. The airport was an old

WW II bomber training base with the typical triangular runway layout. I told John to land short, so we could use the first turn-off. He landed such that the Cub touched down before the runway threshold lights. As I saw the wheels go past them, I yelled: "not that short." A national chain, Executive Air, provided services at the airport. As the Cub stopped at the gas pit, a college age girl ran out and laid down a section of red carpet for John to step onto, which he did with a big grin. The same treatment that the Lear jets get. The Cub was refueled. Then, John and I exchanged duties again with myself as the pilot.

After taking off from Grand Island, I turned to a compass heading of 266 degrees and the Cub was on its way to North Platte, Nebraska--125 miles away. Within the first 30 miles, there were three small towns. The last one, Ravenna, has a large railroad roundhouse that is a perfect checkpoint.

After the roundhouse, the Cub was flying over rolling prairie. Each brown hill, with a wash gully at its bottom, looked just like the last one. To my inexperienced eye it looked as if there were no checkpoints at all. To compound matters, the ceiling was getting lower and without the sun everything appeared to be a shade of slate grey. Twenty miles to our south, John and I could see pillars of white steam from the alfalfa driers in each of the small towns along the Platte River. These pillars reached to the base of the cloud deck. The towns themselves could not be seen as they were down in the river valley. I did not have a map in the front seat and had only a fuzzy idea of where we were. John, on the other hand, knew exactly where we were as he was the navigator on this leg.

After 30 minutes of flying the compass heading, I asked John to calculate a ground speed. As soon as his attention was in the cockpit, I slowly let down until the Cub's wheels were about 2 feet off the ground. After a few seconds of this, his peripheral vision detected something. He looked to the right. Fence posts were going by at eye level. Having his attention, I pulled back up to our usual 100feet. An altitude at which we did not have to worry about the rural telephone and power lines.

To make things interesting, John and I had developed a routine of requiring the pilot to fly the plotted course to the nearest 100feet. When the checkpoints were plentiful, we would chide each other for being off the course more than the allowable 100feet. Then, would insist upon an impossible half degree heading correction or a zigzag that would put us directly over the checkpoint. Being an ex-Air Force navigator, I thought this was great fun.

By the time we landed at North Platte, the weather was deteriorating rapidly. After refueling the Cub and tying it down on the ramp, John and I went into the FSS to check the weather reports. The satellite picture sequence showed that we had spent the early part of the afternoon flying within the confines of the only hole in the cloud cover.

We stayed over night at the Holiday Inn. The next morning John and I awoke to dark low clouds and it was drizzling. With nothing else to do, we walked the four miles back to the airport to check the weather. It did not look good. On the way back to the motel, we stopped at a local store to buy some plastic rain coats. The girl behind the counter looked at us and said

"It's too warm to be walking around town in winter ski jackets."

John replied: "You had better get yours out. It's going to snow tonight."

She said: "That will never happen."

The next morning we awoke to 10 inches of snow on the ground and toward Denver it was reported to be 16 inches deep. We stayed another night. On Sunday, John and I decided that none of the airports between here and Denver would have their runways plowed. In addition, I was running out of time as I had to teach classes on Tuesday. John and I checked on how well the Cub was tied down on the ramp and left it to the elements. We rented a car at the airport and drove though deep snow all the way to Denver. Monday morning, I took the airlines home and my first Cub ferry flight was over.

After a few days, the FBO at the North Platte felt sorry for the freshly restored Cub sitting out in the cold and snow. He put the Cub in their hangar and did not charge us hangar fees. Due to work and the weather, John was not able to get back to North Platte for three weeks. He then flew the Cub to the East Colfax airport where he had a hangar rented for the winter and spring months.

The Return Ferry Flight: Denver to Madison

After what seemed like a very long winter and spring, the Memorial Day weekend finally arrived. I rode the airlines to Denver. The return route of flight was planed to go north out of Denver and through eastern Wyoming to Rapid City, South Dakota. At this point the route turned east and went through central South Dakota, southern Minnesota to Mankato, and then to Madison. The total distance from Denver to Madison was 1,001 miles. John and I called this the "northern" route. The initial section of this route was scenic and took advantage of the generally better spring weather in the west.

This time I brought my movie camera along to record the trip on film. In anticipation of the trip, John had flown the Cub from the East Colfax airport to the Jefferson County Airport (JEFFCO) and tied it down on the ramp. Before dawn the next day, John's wife, Joye, drove us to the airport with two sleepy children curled up in the back seat. Our bags were tucked into the Cub and we said our goodbyes. With a call to the tower, we were on our way to Cheyenne, Wyoming. Takeoff was at 6:45 A.M. on Saturday, May 30,1970 with myself as the pilot and John serving as the navigator.

Northern Route From Denver to Mankato

Initially, a series of front range cities were strung out along our flight path. Near the city of Loveland, the Cub crossed Interstate 25 that then stayed off to our west on its way to Cheyenne. Further to the west, we could see the snow capped Rocky Mountains. To the east was the gently rolling cattle grazing land of Colorado. Although the weather was clear at Denver, about half way to Cheyenne an area of showers developed at the base of the Rocky Mountains. Close to the ground, these clouds formed a dark blue wall. Flying parallel to the line of clouds some light rain was encountered, but the wall of clouds remained to the west of us.

From Denver to Cheyenne the land climbs 500feet in altitude with the latter airport being at 6,156 feet above mean sea level (MSL). The high-pitched Lewis propeller was on the Cub, so climbing was a slow process. The airport was on the north side of the city, so we had to fly across the city at a relatively low altitude. Landing at high altitude airports is different as the stall occurs sooner than you expect and the approach speed needs to be faster than at lower elevations. Fortunately, Cheyenne's runways are very long. I set up final approach for a touchdown point well down the runway and then made a wheel landing.

The Cub was refueled, John and I exchanged roles, and we strapped on the Cub again. As the Cub broke ground, the Cheyenne control tower operator bade us goodby with "The only way to fly." This was Western Airline's current advertizing slogan. John and I got a big kick out of the message.

About 20 miles before the North Platte River, the ground began a discernable descent to the river. Some 30 miles north of the river, we landed for gas at the Lusk, Wyoming airport. Due to the Memorial Day weekend the airport was unattended, so John and I had some trouble getting gas.

Leaving Lusk, the Cub was flown rather high (1,000feet AGL)as we went across the large Angostura reservoir. By now, the weather was clear with only a few small puffy cumulus clouds dotting the sky and the visibility was at least 50 miles. From our lofty vantage point, the reservoir's blue water, the brown buttes around it and the white clouds above it formed a spectacular vista.

When the Cub approached the southern edge of the Black Hills I lost some altitude and deviated from our course, so we could fly past Mount Rushmore. The four faces stood out clearly, but appeared smaller than I thought they would. A low pass was made about a quarter mile from the monument, so John could take some photos of it..

The Rapid City airport's elevation is 3,182feet, which is 3,000feet lower than Cheyenne's. The airport has one very long runway and a control tower. John called in and I landed with no problem. I must be getting the hang of it. We took time to eat lunch at the airline terminal as the food was a lot better than the usual cross-country fare of Stewart's microwave sandwiches.

The next segment was to Pierre, South Dakota 135 miles away on a 71 degree compass heading. The range land is gently rolling with occasional creek beds and country roads. Even though it was spring, the land beneath us was still shedding its winter coat of brown grass. Years later, the movie Dancing With Wolves was filmed here. It was enjoyable to sit in the movie theater and reminisce about flying low over those same hills and valleys.

The city of Pierre sits on the river flats below the east bluffs of the Missouri River. Just north of town, an earthen dam creates a large reservoir called Lake Oahe. The airport, constructed during WW II, is situated to the east of town on top of the tall river bluffs. Pierre did not have a control tower, so all John had to do was fly the pattern and land. The lineboy asked us where we came from. John replied "Rapid City." The lineboy then asked if we had any trouble with hail, I responded that it was clear when we left. Then, the lineboy informed us that a line of thunderstorms had formed over the Black Hills and came rumbling out into the plains about 20 minutes after we had taken off. The line of thunderstorms was accompanied by heavy rain, lighting, thunder, high winds, and large hail. This is a common occurrence in this part of South Dakota.

At the airport weather bureau, a check of the weather showed the line of thunderstorms was headed right at Pierre. John and I decided flying was through for the day, so the Cub was rolled into the old WW II bomber hangar. The hangar was large enough to hold at least one B-17 or B-24 bomber. It was constructed of large wooden beams. The front door consisted of tall metal sections that could be rolled aside. Although most of these hangars were taken down after the war, a few of them survive at former WW II airbases such as Pierre's.

Once the Cub was tucked away, John and I took a taxi to a motel. A few hours later, a much weakened line of thunderstorms came through. As it passed a huge rainbow formed that was unusually brilliant against the background of nearly black clouds.

Sunday morning, John and I awoke to cold grey skies looking as if they would spit rain at any moment. The night before, we had left the Cub in the front row of the hangar. When we arrived this morning, it was behind three other airplanes. How they crammed those other planes in the hangar, we never did figure out. It took a good 20 minutes for John, me, and the lineboy to extricate the Cub from the hangar.

Although the ceiling was only 500feet, the visibility under the clouds was a good 20 miles. The man in the weather bureau said it was below VFR minimums, but John and I thought it was good Cub weather. With John in the back seat as navigator, I took the Cub off and we headed for Huron, South Dakota--a flight of 101 miles on a compass heading of 81 degrees. I immediately pulled my dumb trick of flying towards a prominent landmark not on the course line. After about ten minutes of this, John recognized the error of my ways and gave me a compass correction to get us back on course. My movies of this route segment show that a 30-degree wind correction angle was needed to stay on the course line.

Thirty miles past Pierre, the Cub flew over an automobile accident that had just happened. One car was pretty well smashed and the other was off to the side of the road. People were standing around, but the police had not arrived. John and I were unable to assist as Pierre was out of radio range.

On final approach to Huron, John noticed that the airport rotating beacon was flashing white and green indicating IFR conditions. By this time the Cub was low on fuel and we were committed to landing despite being illegal. To avoid tangling with the North Central Airlines Convair 540's that served Huron a few times per day, I landed on a grass runway perpendicular to the main runway. Then, taxied the Cub right past the FSS to the gas pumps, filled up the tank, taxied past the FSS again and took off.

At this point in time, flying was a lot more casual. Ranchers in this part of the country use airplanes like cars and don't pay a lot of attention to the rules. The FSS staff probably assumed we were one of them. One fellow, I talked with at Pierre, loaded his two children in the back seat of a Super Cub and flew them from his ranch to a country school every day. He landed on the road in front of the school. Everybody concerned thought it was perfectly normal.

From Huron to Tracy, Minnesota there was water standing everywhere. The farmyards were seas of mud and every low spot in a field was a small lake. Farm tractors had driven from the fields onto the highways leaving long trails of mud behind each wheel. At Tracy, the lineboy told us it had been pouring down rain for the past five days. This was the first day it had not rained, but it was still dark and gloomy. Since temperatures were in the mid 40's, John and I were still wearing our down ski jackets, gloves, and hiking boots.

Flying the 83 miles from Tracy to Mankato would take about an hour and twenty minutes. The land was flat until the valley of the Minnesota River was reached. Once there, numerous wooded side valleys merged into the wide main river valley that we followed to Mankato. It was late afternoon when John made a nice landing at Mankato. We stayed overnight with our parents again.

Monday morning, despite the lack of any improvement in the weather, John and I took off for LaCrosse with myself as pilot. We weren't concerned as Highway 14 could be followed for most of the 136 miles. Approaching the small town of Dodge Center, Minnesota, 55 miles east of Mankato, the weather deteriorated. John and I were now faced with a 300foot ceiling, a mile visibility in light rain, and heavy haze. To avoid the infamous radio tower, I positioned the Cub a 100feet above and on the south side of the railroad track. Once over the middle of Dodge Center, I could see the grass runways of their airport. A quick right turn was made and I landed the Cub. No sooner had we climbed out of the Cub than a Piper Comanche appeared out of the gloom flying west at 100feet above the same railroad track. John and I gave each other an "Oh Boy" look that said we had landed in the nick of time.

After waiting in light rain for an hour, it was lunch time. John put on his black bowler hat, I carried the movie camera in its bag, and we walked the several blocks to town. At the local restaurant, the patrons knew we were strangers and looked at us quizzically. Finally, the waitress came over and asked us if we were from Hollywood. I told her that we had just landed at the airport and had walked in to have lunch in her world renown restaurant. This seemed to satisfy her, but she never did make clear as to the origins of her question. John in his bowler hat must have reminded them of Paul Newman in the movie Butch Cassidy and the Sundance Kid.

After lunch, the ceiling was higher, so we took off and easily made it to LaCrosse, Wisconsin. Leaving there, the ceiling was roughly 1,000feet above the floor of the river valley. The visibility was at least 5 miles. Thus, flying was easy, but navigating the 102 miles from LaCrosse to Madison still posed its problems. Coming from the west, I could not find Yuba hiding below the tip of a ridge, but did see Hub City further south in the same valley. Although I was starting to get familiar with some of the landmarks that were exactly on the course line, a half mile or so off the course line was still foreign territory to me. After an hour and a half of flying time, we landed at Morey's and the Cub was mine for the summer. The next day, I drove John to Milwaukee to catch the United flight to Denver.

The Intercom

On the ferry flights to and from Denver, the noise level(92db) in the cockpit was a serious problem. John and I always had great difficulty hearing the little radio's speaker and conversing with each other could be accomplished only by yelling at close range. Once on the ground, both of us had a ringing sound in our ears, and our hearing was impaired. Due to years of flying in cockpits having high noise levels, many old time pilots are nearly deaf. Not wanting to join them, I decided, after the Denver to Madison ferry flight, to install an intercom system. An inexpensive ($6.95) Lafayette Radio audio amplifier on a small printed circuit board, some cheap headsets, and condenser microphones were purchased. Out of all this and some junk box items an intercom was constructed and linked to the handheld radio. The phone jacks were mounted on an aluminum plate, in the wing root area, that covered the hole where the original magneto switch had been. The whole installation was very neat although the headphone cords dangled in the rear pilot's face. Both pilots now had headsets, microphones, and there were push button switches on the top of the control sticks. To transmit via the Regency radio, the person in the back seat still had to lean forward and talk into the radio's speaker. Via the intercom, the front seat person could hear what transpired. The whole system worked very well. The inter-pilot conversations could be carried on in a normal voice and the headsets effectively blocked out the ambient noise. Peace and quiet at last.

Renting a Hangar At Wanuakee

In the fall of 1970, John was not able to find a hangar for the Cub in the Denver area. We decided to leave the Cub in Madison, thus putting the hangar burden on me. About a month later, a flight instructor I knew told me of a hangar for rent at Waunakee. I jumped at the chance as Waunakee was only 20 minutes from home.

The Waunakee airport was founded in 1946 by Jerome Ripp who had converted one of his farm fields into an east-west runway. He received his Private Pilot's license in the same year and has been flying ever since. When I moved the Cub to the airport, he was in the process of developing the land along the runway. There were quite a few houses on the north side and several of them had hangars built into the back of the houses. The lots on the south side were deeper and there was room for separate hangars. There were only three or four houses on the south side.

Perpendicular to the east end of the runway and to the north there were three rows of hangars, a small building, and a gas pit. The first row faced east towards the street. A second row was tucked close behind and faced west. A taxiway separated the second row from the east facing third row. The hangar I rented was in the third row at the north end. The bulk of these hangars had been built in the late 1940's and were becoming rather worn looking. None-the-less, the Cub was out of the wind, rain, and the sun's harmful rays.

Initially, flying out of Waunakee was a bit unnerving. The 2,200 foot main asphalt runway was oriented east and west and had a dip in the middle. A wide grass runway was adjacent to the south side of the main runway and its east end slanted downward to the north. Landing to the west, on the south side grass, the Cub's left tire would touch the ground while the right tire was still about 3" off the ground. As the Cub slowed, the right wing would suddenly drop and the right wheel would touch the ground. It took me the better part of a summer to get used to this and now it seems perfectly normal. Landing to the east, the runway drops away at just about the same rate as the Cub drops in a stall. This can catch me unaware as I am expecting to touch down, but the ground drops away and the Cub is still 2 feet off the ground. If I am lucky, the only result is a small bounce.

One calm day, I was landing the Cub to the east. The Cub was down to about 10feet off the ground when a Piper Comanche, landing to the west, kept coming down on final approach. I decided that its pilot did not see me. I pushed full throttle and turned hard to the left. The Cub skimmed over the rooftops with only 5 feet to spare and not climbing very well. After getting a close look at quite a few rooftops, I finally realized the carburetor heat was still on, turned it off, and the Cub started to climb. I circled around the airport and landed to the west. Later, I talked to the Comanche's driver(he clearly was not a pilot). He said that he saw the Cub, but thought he could land beyond it rather than making a go-around -- a really dumb idea.

Probably, my most "interesting" landing occurred in July. I had been at a Fly-in at Brodhead, Wisconsin when the wind had started to get quite strong. I took off and headed north toward Waunakee. As I went across Lake Mendota, I could see that it was a mass of large, wild whitecaps. There were no boats on the lake. By the time I reached Waunakee, the wind was howling from the southwest at 45 degrees to the runway. I knew I could not land the Cub with that much of a crosswind component. I set up my landing pattern, so the final approach would be directly into the southwest wind. This meant coming in over the tops of the hangars, landing across the runway, and rolling out into a backyard of a house on the south side of the airport. Once the Cub was on the ground, the house could serve as a windbreak.

The wind was so strong that my ground speed on final was less than twenty miles an hour. As the Cub passed ten feet over the hangars, I could see a fellow standing at the door to his hangar. I hoped that he would come over to hold down the Cub once it was on the ground. The Cub touched down on the south grass just past the main runway and rolled uphill into a backyard. I kept the Cub moving until the house broke the force of the wind. Once out of the wind, I opened the door and waved for the fellow to come over. He just stood there. Suddenly, Art Anderson and his son appeared, each grabbed a wing tip. Then, I taxied back to my hangar with them holding the Cub down in the strong wind. Art said that it was one of the most skillful landings he had seen. He was a corporate pilot with many thousands of hours of flying time. I was pleased by his compliment, however, I don't think that I would try such a landing in anything but a Cub.

In February 1971, a snowstorm swept in from the west bringing 6 inches of snow and 40 mph winds -- a classic Midwestern blizzard. The next day I went to the airport to check on the Cub. The strong winds had blown snow through the half-inch cracks between the vertical siding boards. Once inside the hangar, the snow just kept accumulating. The snow was piled up over the wings and the rear of the fuselage could not be seen. It took me the better part of the day to carefully dig away the snow without hurting the airplane. I had visions of a repeat of the situation that damaged the Cub in the first place. Fortunately, the wind was so strong that it blew the snow off the roof.

The Second Madison to Denver Ferry flight

In the summer of 1971, John was able to rent a hangar at the Flying D airport for the upcoming winter. As per our agreement, he would keep the Cub for another winter season. Thus, another round of ferry flights were in the offing. John and I thought an earlier departure might avoid the snow of the previous trip. Hence, the ferry flight to Denver was scheduled for the last week of September. Joye recently had received her Private Pilot's license and in the keeping with family tradition would immediately go on a long cross-country flight. John and Joye would fly their Cessna 182 to Mankato. I would fly the Cub there via LaCrosse. Because the Cub was now based at Waunakee, there was a new course line to LaCrosse. This meant I had to learn a new set of checkpoints.

I took off at 8:00 A.M. on Tuesday September 28,1971. The Cub crossed the Wisconsin River at Sauk City and in a few minutes went over the Sauk Prairie airport. Directly ahead on the course was a saddle in the west end of the 500foot tall Baraboo Bluffs. In the middle of this saddle was an abandoned farm house whose paint had disappeared long ago and the weather beaten wood siding was a dark shade of grey. The next checkpoint was the aptly named hamlet of Hill Point located at the bottom end of a sharp ridge.

Since it was fall again, the farm fields in the valleys and on top of the ridges were a golden tan color. The leaves of the hardwood trees had turned. The reddish brown leaves of the Oaks, the yellow of the Birches, and the reds and yellows of the Maples created a montage that looked like a multi-hued carpet.

The next checkpoint was the hamlet of Rockton in the valley of the Kickapoo River some four and one half miles up river from LaFarge. Because the river would flood the towns downstream from Rockton, the Army Corps of Engineers started constructing a flood control project that would put a dam on the Kickapoo River. At Rockton, the road that paralleled the river was relocated to higher land on the east side of the valley. A new concrete bridge was put across the river at Rockton. Due to political pressure brought by the residents of the Kickapoo valley, the dam was never built. As a result, the yearly floods continued.

There is a readily identifiable sharp end of a ridge just south of Rockton. I reached it right on my ETA. A few miles west of Rockton, the church at Newry can be seen. When, the ski jump at Westby could be seen standing out against the horizon, the Cub was almost on the earlier course line. From this point to LaCrosse, the valleys of the un-glaciated area are much deeper and more densely wooded that they were further east. At St.Joseph, I deviated north into the valley of the LaCrosse River again to get a better view of the airplanes that might be in the LaCrosse airport's traffic pattern.

On the LaCrosse to Mankato leg I stayed on the plotted course line. After St. Charles, the gasoline storage tanks just north of Eyota could be seen and Rochester is only a few miles further. In another 45 minutes, the Cub flew just north of the Waseca airport. No need to climb up this time to clear the town. After two hours and twenty minutes, Mankato came into view and I landed at the airport. John and Joye's 182 was parked on the ramp.

My mother was 68 at the time and over my father's strong objections she decided to go with us to Denver. My mother was the adventurous one of that pair. While living with her sister and working as a "soda jerk" in Billings, Montana she met Charles Lindbergh, who was a barnstormer at the time. She parted with some dollars for an airplane ride with him. In the early evening, she and her girlfriends went to the airport, pulled off the cockpit covers, and each took turns sitting in the rear cockpit of his airplane. Unfortunately, the photographs of my mother sitting in the airplane were taken quite close and I could not determine the type of airplane. It looked as if it might have been a Lincoln-Page biplane or possibly a Swallow. We also have a photograph of her sitting on the engine cowl with her legs resting on the prop. All her life she told the story of her meeting Lindbergh.

At 7:00 A.M., on the morning of September 29, 1971, the two planes departed Mankato. Joye was in the back seat of the Cub serving as navigator and I was the pilot. Mother (we called her "Granny") was in the 182 with John. The flight plan was to follow the "southern route" to Denver.

There was a thin layer of ground fog on the corn fields and in the woods. The rising sun peeked through at a low angle backlighting each patch of fog and glistening the dew on the ground. Looking toward the sun, Joye and I could see multiple layers of fog each with a different golden hue giving the world around us a surreal look.

This time I was prepared for the three towns on the ridge, ignored them, flew a compass heading, and stayed on course. We landed at Spencer to refuel and then Granny replaced Joye in the Cub. She has always been very agile and had no trouble getting into the front seat. Ordinarily, I put the passenger in the rear seat, but to land at Sioux City I needed to be in the backseat to use the radio. After takeoff, Granny took a great interest in the countryside and watched intently as the farm fields slowly passed beneath us.

After refueling the Cub at Sioux City, a call to ground control was made and I taxied the Cub to the active runway. With another call to the tower, Granny and I were on our way to Columbus, Nebraska. I am always happy to leave an airport with a control tower as I am never sure that I am doing things correctly or not.

At Columbus, Granny went back into the 182 with John for the leg to Grand Island. Joye climbed into the back seat of the Cub to be the navigator again. The weather was cooperating and so far it had been sunny. After taking off, we followed the railroad and highway that parallel the Platte River from Columbus to Grand Island. The river has numerous small branches that weave back and forth in the river channel. By this time of the year, there was not much water in the river. We were flying the Cub just above the tops of the telegraph poles. Joye took some movies of the semi-trucks, on the adjacent highway, that were passing us. Off in the distance, I could see the next small farm town where there was a tall wooden grain elevator to the left of the railroad tracks. As the Cub approached the town, I looked for wires from the grain elevator across the railroad tracks. Not seeing any, I held my altitude. The Cub flew past the grain elevator just above the second story windows. Joye took movies of the whole event. Years later, I saw the movie The Great Waldo Pepper where a Standard J-1 biplane, with a girl hanging onto a wing strut, was flown down the main street of a town in Texas. I knew just how they must have felt.

The airport at Grand Island is another of the former WW II airbases. It has the usual triangular set of runways that were still in usable condition. None of the WW II buildings remained. The Cub was landed at Grand Island, but this time we did not get the red carpet treatment.

Joye stayed with me for the leg to North Platte. This was my first time flying this 125 mile leg without John's guidance. Leaving Grand Island, on a heading of 266 degrees, the three villages were the key to getting set up on the course line. By the time we reached Ravenna and its railroad roundhouse, Joye had established a good wind correction angle. Once past Ravenna, the prairie is intersected with gullies that are about 30-50 feet deep with numerous side gullies feeding into them. Short, bushy, dark green trees grow in the gullies, but not on top of the ridges. This time, the flight over the desolate country between Grand Island and North Platte was a lot more comfortable for me. Joye's map reading skills kept us right on course for the whole flight. I was seeing some of the same hills, roads, and clumps of trees that I had seen two years earlier. Even at the low altitudes the Cub flies, I got a sense that the ground was rising as we flew to the west. After two hours and twenty minutes of flying time, Joye and I landed at North Platte where John and Granny were waiting for us. The four of us stayed over night at the Holiday Inn.

Thursday morning's weather was sunny and nice, so it was Granny's turn to ride with Joye in the 182. Once they were in the 182, Joye informed Granny that she was her first passenger. The news didn't bother Granny a bit, she figured that if Joye had a license, she was a good pilot. John and I were in the Cub. The route of flight from North Platte to Julesburg, on a compass heading of 252 degrees, was never more than six miles from the South Platte River. There were seven small towns strung out along a two lane highway on the north side of the river. Even though they were not exactly on the route of flight, these towns made excellent checkpoints. As the Cub approached Julesburg, Colorado, the weather began to deteriorate. The ceiling was getting lower, the sun had disappeared, and light rain had started to fall. John and I landed and in a few minutes, Joye and Granny arrived.

Since the weather was getting worse, John would take Granny with him in the 182 while Joye and I would fly the Cub. We would regroup at the Fort Morgan, Colorado airport ninety miles to the southwest. By this time it was raining harder, but the visibility was not all that bad. Joye and I took off with myself as pilot. As the Cub crossed the highway we could see that the cars coming from the south had their headlights on--not a good sign. Turning onto the course heading of 220 degrees, I flew a couple of a hundred feet off the ground in heavy rain. The radio was on and a Cessna 172 flying north was heard calling into Julesburg. Joye gave a call on the radio, but received no reply. A few seconds later we saw the plane about 50 feet above us and a city block off to the left. I'm sure the pilot never saw us (you are never alone!)

Because of the rain and limited visibility, Joye and I were kept very busy. I was concentrating on keeping the Cub about 200 feet above the ground and holding the compass heading. Joye was getting only fleeting glimpses of the country roads and other landmarks. As a result, it was difficult to match them with what was on the map. The Cub flew only a mile or so to the east of Sterling, but we never saw the town.

Ten miles before Fort Morgan, the Cub flew out into clear skies. The airport is located five miles north of the town, has a northwest-southeast hard surfaced runway, and a lone building. From the air, it looks like it is in the middle of nowhere. After landing, Joye and I looked back at a huge towering Cumulus that must have gone up to 50,000feet. Underneath the cloud, it was pitch black. Surprisingly, we had encountered no turbulence and only light winds under this Cumulonimbus. The Cub was refueled. Then Joye and I waited and waited, but no John and Granny. Finally, we saw the 182 coming in from the east rather than from the north.

After he landed, John told us what had transpired. When they took off to the east, very heavy rain was encountered. John could not see anything out the front and only a little straight down. To compound matters, water started leaking in and fogging the windshield. Granny took out a handkerchief and started wiping off the water so John could see out. As they crossed the highway, he turned left, followed the highway until he saw the road sign pointing into Julesburg and made another left turn. Upon reaching the main street, a left turn to the south was made at tree top level. In a second or so, he could see the International Harvester dealership. He knew it was just off the southwest end of the main runway. John pulled a tight left turn, chopped the power, and landed on the runway without even getting a good look at it. Granny's only comment was: "John, have we changed course?" From then on, whenever John and I flew into really bad weather, she was quoted verbatim. At

Julesburg, Granny and John waited for the weather to let up some, took off, then flew due east toward Akron, Colorado until they were out from under the thunderstorm. Then they flew around it to Fort Morgan.

The rest of the trip to the JEFFCO airport was in typical Colorado severe clear with no wind. Another ferry flight was history. I rode the airlines home the next day. Granny stayed on a few more days before she went home on the airlines. A few days later, John moved the Cub to its winter home at the Flying D airport.

Return Flight: Denver to Madison, May 1972

The return trip to Madison was scheduled for the Memorial Day weekend of 1972. A few days before then, the Cub suffered a minor accident that looked like it might delay the trip. On a local flight, John had just touched down in the Cub, when the left landing gear folded upwards, and the left wing tip touched the ground. The Cub slowly turned off the runway and onto the dirt. Fortunately, the prop stayed well above the ground. The lower left shock strut had broken a few inches from the top. This allowed the landing gear leg to rotate upwards. The local mechanic straightened the bent rear landing gear leg, laid some fabric over it and the scrapped wing tip. He then doped the fabric up through silver. Later, a metallurgist at the University inspected the shock strut. He said it had been bent at some time. Subsequent landing stresses had caused it to work and finally fail at the bend. John and I thought that the shock strut was bent the time the Cub rode out 80 mph winds at JEFFCO while chained to a ground cable.

This crisis resolved, we decided to take the southern route back to Mankato and then to Madison. The overall plan was to fly both the Cub and the Cessna 182. I would stay in the Cub while John and Joye would alternate between the two airplanes.

The two planes departed JEFFCO early on Saturday May 27,1972. The trip to North Platte, with a gas stop at Sterling, Colorado was uneventful under clear skies. As the route was a familiar one for John, he had no problem staying on course while I navigated. From Sterling to North Plate, I could see Interstate76 and the South Platte River for the whole flight segment. At each stop, Joye in the 182 would wait about ten minutes before taking off, pass us between airports and be waiting for us when we got there.

After North Platte, John and I headed the Cub for a gas stop at Grand Island 125 miles away on a heading of 86 degrees. Even though there are not many landmarks. This leg also was becoming familiar: recognizable wash gullies, edges of hills, small creeks and an occasional ranch were appearing right on time. Of course, little of this was on the map. In that it was springtime, the color of the range grass varied from light to medium green with an occasional small area of bright yellowish green. In a little over two hours, the Grand Island airport was reached and we made a fuel stop there.

On the leg to Columbus, Joye and I were in the Cub. Although the weather had been perfect up to Grand Island, the Cub was beginning to catch up with the back side of a line of thunderstorms. Several miles ahead of us was a wall of tall grey clouds with showers pouring out of their dark bottoms. Yet, under the clouds there were breaks in the line of showers where you could see the distant sky through the gaps. Joye and I decided to fly through one of the more promising looking gaps. Once under the clouds, at about 200feet, it was quite smooth. In some 10 minutes, we were getting to the front side of the line of storms. As we broke out from under the clouds, I looked up. Just above the Cub was a dark grey, cigar-shaped roll cloud that precedes a thunderstorm. I quickly grabbing my camera and took a picture of it. Such clouds are noted for their accompanying turbulence, but it was smooth flying all the way under it.

In Columbus, the FSS briefer asked: “Were, did you come from?”

I replied: “Grand Island.”

He was quite surprised as the radar summary showed a line of thunderstorms right across the course line.

"How was the weather?"

"It was very smooth with good visibility."

I don't think he believed me, however, he neglected to ask me how high we were flying.

After refueling in Columbus, Joye and I took off and headed the Cub towards, Norfolk, Nebraska-- a town not on the original "southern" route. Once there, John and Joye swapped planes again. John and I headed for Sioux City, Iowa. By now it was getting late in the day and the checkpoints were not coming up properly. John, the navigator, spotted a town's water tower some two miles away. A quick buzz job allowed us to read the name of the town. Checking the map, we found that the Cub was seven miles off course. So much for the 100 foot rule. By the time Cub touched down at Sioux City, twilight was approaching, it was beginning to rain, and the wind was picking up. As I taxied the Cub to the ramp, the 182 was not in sight.

The Cub was put in a hangar and we waited for Joye to arrive. Forty-five minutes later--still no Joye. Checking the map, John located an airport at Wayne, Nebraska that was slightly off the course line. John called there. The fellow answering the phone said that Joye had landed, waited a while, and then taken off again. It kept getting darker and darker with lightning flashing here and there. Our tame line of thunderstorms had become rather ferocious looking.

Just as the last light of day disappeared behind the wall of weather, the wing tip lights of the 182 appeared out of the gloom -- Joye had arrived. John and I decided that she had the makings of a real pilot. She told us that she had waited a while at Norfolk and once airborne ran into rain just past Wayne. She could not work around it, so she flew back to Wayne and waited there until that cell had moved out of the area before proceeding. She was very calm and matter of fact about her flight, while John and I had been quite concerned about her safety.

Later, Joye would become very active in the Denver chapter of a woman pilot's organization called the "99's". She would often ride right seat on a cross-country race that they had sponsored. When things started to look soupy, the pilot would want to turn back. Joye would lead her through the weather without getting in trouble or lost. A feat the impressed the other 99's. She told them that this skill was developed on the Cub ferry flights. Within a few years, Joye also earned a commercial pilot's certificate, an instrument flight rating and took aerobatic training. In 2004, the Midwest Antique Airplane Club presented her with the "Lady Antiquer of the Year" award.

After remaining over night (RON) at a local motel, the next day(Sunday) dawned with a solid overcast at 800feet with a five-mile visibility under it. Since the Sioux City airport is in the Missouri River valley, the clearance between the tops of the river bluffs and the clouds was only a few hundred feet. The control tower operator gave both planes a special VFR clearance, so we could takeoff. He also warned us about the very tall radio towers on the bluffs right on our course to Worthington, Minnesota --95 miles away on a heading of 20 degrees. As a precaution, Joye and I, in the Cub, followed a highway through town and up the bluff. The Cub passed within a city block of one of the towers, whose upper two thirds were in the clouds. Joye and I worried about how John was going to get the 182 through there. Once on top of the bluffs, the visibility was good and after 20 miles or so the ceiling lifted somewhat.

Before leaving Sioux City, Joye and I had arranged to meet John at Orange City, Iowa to talk things over. Joye and I landed there and in a few minutes, the 182 landed. Although it was a cloudy grey day, the visibility was good, so we agreed to keep flying. Both planes took off for Worthington, Minnesota. In the fields below us, farmers were planting corn and soybeans in the freshly plowed and harrowed fields. About halfway there, John dropped some flaps on the Cessna and came down to fly formation with the Cub. As the Cub rose up over a small hill, Joye saw a set of power lines directly ahead of us. She called John on the radio to warn him, but received no reply. Joye and I watched with some trepidation as the 182 flew by only a few feet above the wires. John never saw them.

At Worthington, the sky remained overcast, but the ceiling was lowering. It still looked like good Cub weather, so Joye and I took off and headed for Mankato--91 miles away on a compass heading of 61 degrees. A railroad and a highway going to Mankato were a few miles to the north of the course line. Just before St. James, Minnesota, a thunderstorm popped up and it started to rain hard. Joye and I decided it would be prudent to be on the ground, so we landed at this town's small unattended airport.

As John came across town in the 182, at tree top level, he had to lift a wing to clear a small radio tower on top of the local sheriff's office. That considerably dimmed his enthusiasm for flying a fast airplane through heavy rain at a low altitude. John and I stood under the wing and Joye sat in the Cub while three or four small thunderstorm cells went through. Heavy rain accompanied each one of the cells, but each cloudburst was short lived. After an hour of this, the last one worked its way toward the northeast and the weather was flyable again. It was only a half hour flight to Mankato. Through the Cub's skylight, I could see John flying the 182 high above us the whole way.

We stayed with our parents, who by now were getting accustomed to wayward cross-country pilots dropping in on them without any warning. By the second night, our traveling weather companion finally moved off far enough to the east that it was no longer a factor. In the morning, we awoke to clear blue skies, nice temperatures, and little wind. For the rest of the trip, Joye took the 182 while John and I flew the Cub.

On the way to LaCrosse, Wisconsin, John, in the front seat, was contour flying the valleys that merge into the Root River valley. As we came to the west side of a valley, John would dive the Cub and then fly low across the valley. On the east side he would climb the Cub up out of the valley. Since these valleys are close together, a roller-coaster like ride was produced. When the Cub popped up out of one of these valleys, there was a farmer standing alongside of his car relieving himself onto the gravel road. All of a sudden he was face to face with a yellow airplane. John gave him a wing waggle and I smiled and waved as we passed about a wing's length away from him at almost eye level. John and I got a good chuckle out of that one. I've often wondered what the farmer's reaction was, but I'll bet it was better than that of a Russian seaman my crew did a number on in Korea.

One of our over water weather reconnaissance routes went near Port Arthur, Manchuria where there were many merchant ships, mostly British, carrying goods to the Chinese. On these long over water flights, the B-26's carried a large teardrop shaped gasoline tank under each wing. To stay under the Chinese radar, these flights were flown at about 20 feet off the water. On one of these flights, my pilot spotted a merchant ship on the horizon. He decided to fly toward it so the weatherman, in the clear glass nose, could use the hand held camera to take its picture. As we approached, I could see a seaman standing on the bridge of a Russian ship whose deck was loaded with war materials. He either heard or saw us and quickly realized he was looking straight at the nose of an American bomber with what looked like a torpedo under each wing. As we bore down on him, he thought we were not going to clear the ship. He jumped off the bridge and landed in a lump on the deck some 20 feet below. As we pulled up over the ship's mast, the radio man in the rear compartment told us that the seaman did not get up. One for our side!

From LaCrosse to Madison, I was the pilot. After the St. Joseph's convent passed under the wing, John folded up his maps assuming that I knew the rest of the 90 miles home. Surprisingly, I stayed on course. At about 50 miles from Madison the Baraboo Bluffs could be seen to the left of course. A high hill, called Blue Mounds, was 40 miles off to our right. All I had to do was to stay to the south of the Bluffs and let Blue Mounds appear to move slowly to the west. Eventually, the Capitol building in Madison appeared on the horizon. This leg was becoming "a piece of cake." After landing the Cub at Waunakee, it was refueled, put in its hangar, and was mine for another summer. Joye had flown the 182 to Morey airport where it was tied down. After resting for a day, John and Joye flew the 182 back to Denver.

For me, sharing the Cub was an ideal arrangement. Ferrying the Cub to and from Denver, allowed me to learn about cross-country flying under John's tutelage. The Cub was flown for hours, at low level, over prairie that best can be described as endless. In addition, we flew through all kinds of weather and generally stayed our course lines. From this I relearned the importance of flight planning and staying on course. Most important of all, I learned what kinds of limitations the weather imposes on flying the Cub.

John and I talked about ferrying the Cub to Denver again, but it never transpired. John usually had two or three airplanes on hand. As a result, his interest in the Cub waned, so I bought John's half interest in the Cub. Now that I was the sole owner of the Cub, I realized that a more secure hangar arrangement was needed as the fellow who owned the hangar I was renting needed to use it again. I got Jerome Ripp's approval to built a hangar at the Waunakee airport. I drew up a set of plans for a T hangar. The construction began when the eight main hangar posts were installed in the ground and the 40 foot front beam was lifted into place. Then, my son, Chris, and I put up the center cross beam, the rafter's, and the one by eight cross boards. My daughter, Ann, cut what seemed like a thousand sections of one by twos to fill the gaps between the cross boards on top of the rafters so birds would not have a place to roost. Finally Chris and I sheathed three sides and the roof with corrugated sheet metal. No door was installed as the east facing front of the hangar was left open. When completed, the hangar was the nicest one in the three rows. Chris, Ann and I were quite proud of our handiwork. Now, I owned the Cub, the hangar was on safe ground, and I was the newest member of the Waunakee airport crowd.

Chapter 4
To the East Coast With Ann

In the spring of her senior year of High School, my daughter, Ann, started taking flying lessons at Morey's airport. After a few weeks, the flight instructor indicated she would be making her first solo flight. On the appointed day, the whole family was at the airport to witness the event. The skies were clear, but there was a strong gusty wind from the south. Ann made one landing with her instructor in the plane, then he climbed out, and let her proceed alone. Ann handled the Cessna 150 like a pro. Even though the wind was buffeting the plane on final approach to runway 18, she made three very nice landings. She had soloed after only five hours of dual instruction, beating her dad's record by thirty minutes. The next task was for her to learn how to fly the Cub. The transition from a tricycle landing gear airplane to one with a tail wheel is a difficult one for many pilots. Ann's instructor let her go solo in the Cub after only an hour and a half of dual.

She quickly built up her flying time and was ready for her long distance solo cross-country flight in the Cub. It would be to La Crosse and then to Lake Elmo, Minnesota. She would stay overnight there with her cousins and then return the next day.

Two hours after she left Madison, the phone rang and Ann had quite a story to tell. When the La Crosse airport was within sight, she called the FSS on the Cub's Regency radio, but received no answer. After calling a few more times she called in on the emergency frequency again with no response. Finally, she decided to land. On the roll-out, after the wheels touched the ground, the nose swung to one side. Ann was a bit late on the rudder and she ground-looped the Cub. In the midst of the action the engine quit, leaving her sitting in the middle of the runway. She tried to restart the engine, but it would not start. When another airplane was seen on its final approach, Ann pulled the Cub onto the grass alongside the runway. Seeing her predicament, two mechanics from Viking Aviation came out and moved the plane back to the ramp. Upon inspecting the engine, the mechanics discovered two sparkplug wires that were only finger-tight. I had cleaned the sparkplug's a few days before and had missed tightening two of the leads. Ann was not too happy with me, but I didn't think that the leads were the cause of the starting problem. After refueling the Cub, Ann climbed in, the mechanics started the engine, and the rest of the trip was anticlimactic.

Later in the summer, she took a three-day course from a company specializing in FAA exams. The next day, she took the Private Pilot written test. Ann's score on the test was higher than mine and her instructor's, which pleased her immensely. A week later, she took her check ride and was awarded her Private Pilot's license.

In the fall of 1970, Ann went off to Prescott College in Arizona. Once at school, she went to the local airport and told them she wanted to rent an airplane. They only had low wing Piper Cherokee 140's. Ann and an instructor climbed in and took off. After a few minutes of air work and a landing, he let her go solo. When she told me this, I was amazed as she had never flown a low wing airplane and had no more than 20 minutes in this one before going solo. This was the beginning of our motto "Piper Cub pilots can fly anything!".

When Ann came home from Prescott College, at the end of her freshman year, she and I decided to fly the Cub to Washington, D.C.. Ann could visit a college friend of hers and I would do some consulting. The route of flight was south to DeKalb, Illinois and then east to Morgantown, West Virginia and then into the D.C. area--a distance of 746 miles.

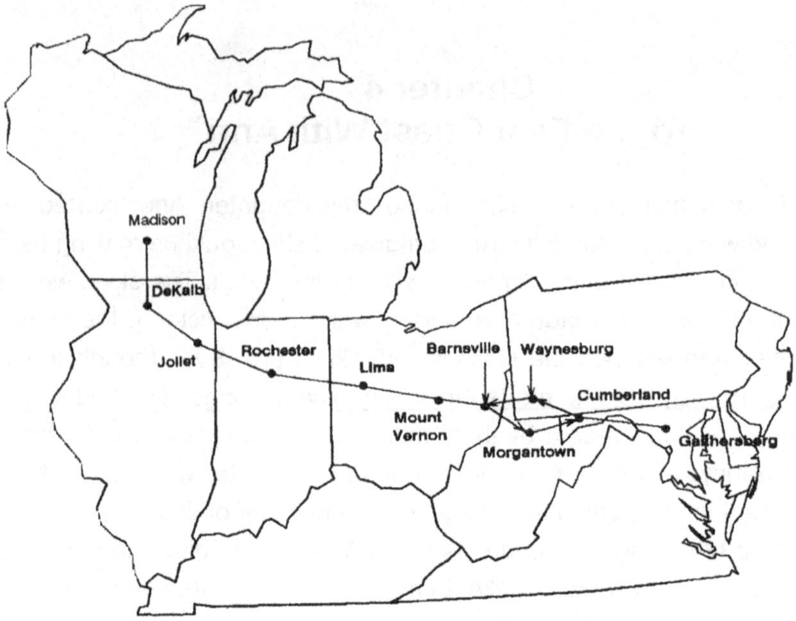

The Route of Flight Madison to Washington D.C.

Ann and I left early on Tuesday, June 8, 1971 with myself in the front seat and Ann in the back seat serving as navigator. After takeoff, a left turn put us on a southerly heading across the City of Madison. The State Capitol building is located on an isthmus separating Lake Mendota from its smaller southern neighbor, Lake Monona. The white marble capitol dome is of the same design as that in Washington D.C. In a few minutes, the Cub was at the southern edge of the Madison suburbs. Then, we took up a south-southeast heading across the farmland toward DeKalb, Illinois.

The DeKalb airport has a single long asphalt runway. We landed to the west, taxied to the fixed base operation at the far west end of the runway, and refueled the Cub. Taking off to the west was a bit unnerving as the runway was terminated by a large hangar and industrial buildings. The Cub's low rate of climb gave us considerable opportunity to look at the roofs of these buildings. The southerly course took us over flat farmland devoted mostly to corn. At the town of Sandwich, Illinois, a more easterly heading pointed the Cub towards Joliet, Illinois that was our next gas stop.

Joliet, has a nice airport with 80 octane gasoline and two runways. Since Joliet is at the outer fringes of the Chicago metropolitan area, the visibility was poor due to heavy air pollution. I chose to land the Cub on the east-west grass runway rather than on the asphalt runway.

Right after take off from the Joliet airport, the Illinois River was crossed where there is a very large coal fired electrical generation plant with three 450foot tall smoke stacks. Ann and I got a close-up look at the plumes of white smoke flowing upward from the power plant's smoke stacks. The abandoned Fort Sheridan army ammunition depot could be seen just south of our position. On this leg, the Cub was flown somewhat higher to avoid the large number of power lines that all seemed to flow towards the Chicago metropolitan area.

Now, the countryside was flat with numerous roads, highways and small towns that made it easy for us to stay on course. We landed at the Rochester, Indiana airport and refueled the Cub. I had been in the front seat ever since Madison and my legs were beginning to feel cramped. Ann and I switched seats, so I could have the luxury(?) of the backseat.

In eastern Indiana and western Ohio, there are a number of Amish communities. Ann would let me know whenever she saw a horse drawn buggy on the road below. Coming into Lima, Ohio, we flew quite close to an oil refinery having a tall pipe topped with a huge flame. In the fabric covered Cub, I felt somewhat like a moth being attracted to the flame. At the Lima airport, we talked to a fellow from St. Louis, Missouri flying a nice Luscombe Observer. His wife sat in the rear facing observer's seat and enjoyed the view as they flew along.

The leg from Lima to Mount Vernon, Ohio was 83 miles long on the same 110 degree compass heading that we had been holding since Joliet. There was a large power line that closely paralleled our course line and facilitated our staying on course. At Mount Vernon I landed long. As a result, I thought we were going to run out of runway. When I gave the brakes a stab, the brakes grabbed. The tail of Cub started up and the nose went down. I gave the stick a good tug backwards and got off the brakes before the prop could touch the runway. Cubs are notorious for having nonexistent brakes. In the past, these had never done much toward slowing the plane down, so I was surprised when they grabbed.

On the flight from Mt. Vernon to Barnesville, Ohio, the Cub started flying over hilly countryside with coal strip-mines that utterly devastated the land. Down in the mines, huge trucks were loaded with coal. Then, they slowly drove up and out of the pits. At one point, Ann and I flew past the largest steam shovel in the world. By now it was late afternoon, the sun had disappeared behind a dark cloud layer, and it had become a dreary day. The Barnesville airport was on top of a ridge. After refueling there, it was on to Morgentown, West Virginia only 77 miles away.

In 20 minutes, The Cub flew across the Ohio river just south of Wheeling, West Virginia. I made a mental note of two tall radio towers, one on each side of the river. Down in the river valley, there were a number of large industrial plants alongside the river and a string of grimy looking cities. Once across the river, some altitude was gained so that we could see over the ridges. Ann and I flew past Waynesburg, Pennsylvania where we abandoned our100 degree heading and started following a highway going directly to Morgantown.

The Monongahela River splits the city. The University of West Virginia is located on the east side of the wide river valley. We landed the Cub at the Morgantown airport just before sunset. Checking the logs, Ann and I had flown the 571 miles from Madison to Morgantown in eight hours and 40 minutes. As there was very little wind most of the way, the average ground speed was 66 mph.

Ann and I gathered up our belongings and took a cab to a motel. The next stop was the home of a former UW graduate student, who had been in my statistics class at one time. He also had been a ski instructor while in Madison. His wife was not expecting anybody and greeted a couple of disheveled looking pilots as if we were people that she did not really want to know. Later, he showed up at the motel and we visited for several hours. The next morning, he drove us to the airport.

Morgantown had scheduled airline service and the local FSS acted as an unofficial control tower. Ann gave the FSS a call and we took off to the east. In the Appalachian mountains, the ridges run in a general southwest to northeast direction and are nearly completely wooded. Here and there would be a small farm on a ridge top. An occasional home would be alongside a creek in a valley. Since this was the first time Ann or I had flown over mountainous terrain, the Cub was flown quite high (1,000feet AGL) to give us some glide time if something went wrong. The haze was cutting the visibility to 5 miles and a large reservoir suddenly appeared out of the haze. A quick check of the map revealed that the Cub was five miles south of course. A heading correction was made that brought us to a branch of the Potomac River that was followed to Cumberland, Maryland. From our vantage point, Ann and I could see the famous Cumberland Gap that provided the early settlers a route through the mountains to Kentucky.

The airport sat on a plateau at a big bend in the river. A small cropduster airport was below the southeast end of the main runway. We didn't see any activity down there, but it must have been interesting during the growing season. We landed at 10:07 A.M., refueled the Cub, and then took off as soon as possible.

To the east of Cumberland, the ridges were lower, less wooded, and eventually the countryside became rolling farm country. Ann and I flew past an isolated peak known as Sugar Loaf Mountain. In the late 1960's, a TWA 727 had crashed into it after being cleared to a low altitude by air traffic control(ATC). The pilot thought the air traffic controller would not clear him to an altitude below the highest obstruction, but he was wrong. Ann and I landed at the Montgomery County Airport near Gaithersburg, Maryland. I rented a car and obtained a room at a local motel. Ann's girl friend came out and picked her up.

Thursday, I went to consult for my friend Morey Cove. We had met when I worked on the Polaris missile project at Control Data Corporation. Friday evening, his family and I went out to the airport for Cub rides. His wife, Betty, really enjoyed the Cub ride and didn't bat an eyelash when I opened up the side door. When I took Morey up, I never closed the side door so that he could watch the ground slip away as we took off. He knew he had been "had" and did not say a word about the door. Several months later, he mentioned that it took him aback at first, but as soon as we were several hundred feet up in the air it was a lot of fun.

Saturday morning, June, 12 1971, Ann showed up right on time. We departed, on a cold gloomy day, under a 5,000foot ceiling with three mile visibility in heavy haze. As the Cub flew on a westerly heading into the Appalachian Mountains, patches of clouds started to appear in the valleys, but the tops of the ridges were above the clouds.

In the mid 1920's airmail was flown across these mountains in biplanes that did not have much more instrumentation than does the Cub. On many flights, these airplanes flew into weather much like that Ann and I were encountering today. The weather often deteriorated rapidly and many airmail planes made forced landings, some flew into the mountain sides, and often the pilots did not survive. Because of this, the Appalachian Mountains became known as "the graveyard of airplanes". Being aware of this, Ann and I were keeping a sharp lookout for places to land if the weather suddenly went bad.

We had no problem flying back to Cumberland and stopped there for fuel. Ann took the front seat. As we taxied out, a Cherokee 140 could be seen shooting touch and goes. I let the Cherokee go by and then took off behind it. When the Cub was about 50feet in the air, a cloud suddenly formed and we were inside it. I could see downward reasonably well, but all I could see forward was grey fog. I knew by now we were probably off the southeast end of the runway. In a few seconds the Cub flew out of the cloud. It was the only one in the valley. I think somebody was trying to tell us something, but the weather didn't look too bad. The ridges to the west of Cumberland topped out at about 3,000feet, so we had to climb as quickly as possible in order to clear them. Because the visibility was deteriorating, Ann and I decided to go up to 4,500feet to give us a somewhat better overall view of the landscape.

By the time the Cub approached Morgantown, several layers of broken clouds had formed between us and the ground. Before leaving Gaithersburg, I had filed a flight plan. By now, it was obvious that Morgantown was not going to be VFR. In addition, the airlines would be flying IFR approaches to the field. My calculations showed that we were only a few miles to the northeast of the airport. Our marginal Regency radio was used to call the Morgantown FSS. The flight plan was canceled. Then, I told them we were going to Washington, Pennsylvania.

At this time, I thought we would be better off under the clouds than above four different layers. Each layer was broken so we occasionally could see the ground. Ann was in the front seat and could see better than I could from the back. I told her to take us down through the clouds. I was totally unprepared for her fighter pilot instincts. The minute I gave her the controls, she whipped the Cub into a steep right-hand diving turn and went down through a hole in the highest layer of clouds. Once under it, Ann pulled a swooping turn around the edge of a cloud bank, quickly found another hole, rolled left, and dove through it. She leveled off just long enough to get the lay of the clouds, made a couple of turns to get positioned, and then made another diving turn through a hole. After about three minutes of this, she leveled off under the clouds at 200 feet above the ground. She turned around with a big grin and said "it's all yours."

The weather was getting worse, so we landed at Waynesburg, Pennsylvania that had been overflown on the way out. Ann and I spent an hour on the ground hoping the weather would improve. One of the local sights was the most awful looking homebuilt airplane that I had ever seen. It was very dirty aerodynamically and looked like it was very heavy. The locals said that it had never been flown.

It appeared as if the weather was not going to change a lot, so Ann and I decided to go to Barnesville, Ohio. I was in the front seat as pilot and Ann was in back to do the navigating. As the Cub climbed to the west, we went across town and then over a ridge. About 15 minutes after takeoff, the visibility was down to less than a half mile. I was flying the Cub at 50 feet above the tree tops when we flew into a fog bank. I lost sight of the trees and was flying totally blind in the fog. I knew if I tried to turn, a wing tip could easily snag the tree tops. I yelled to Ann "I cannot see the trees." She leaned forward and said very calmly "don't panic."

A number of thoughts raced through my mind. Foremost was that even experienced IFR pilots can hold a non-instrumented airplane level for only three minutes before going into a spiral dive. My total IFR time was only an hour, so holding it level for very long was not very likely. I knew the Cub was trimmed to fly reasonably well hands off. Our only chance was to just hold the stick and rudder pedals where they were. All I could do was hope that the Cub had more sense than I did.

After what seemed like an eternity (it was probably 45 seconds), the Cub popped out of the fog bank. Now, the visibility was up to three fourth's of a mile, but I was still concerned about the two radio towers near Wheeling. In another ten minutes we were at the Ohio River and right on course. We never did see the two towers. By the time the Cub arrived at Barnesville, the weather was relatively good. Ann and I had just flown through a front pushing a mass of moist air in front of it. I was thankful we were able to fly through the fog without crashing. I also learned that Ann was a really cool customer in a very dangerous situation.

Leaving Barnesville, the clouds now were behind us and we were flying directly into the afternoon sun. Since we were retracing the same course, the countryside beneath us was familiar. Ann and I relaxed while slowly moving across the strip-mining country. Both of us still had on our warm clothes and jackets that were needed while flying over the mountains. Further west, I noticed that Ann's head was bobbing a lot and that she did not seem to be responding to my queries. I finally shook her and found that she was nearly passed out from the sun beating in on her in the front seat.

In order to get some air going through the cockpit, I opened up both side windows. The air helped a bit, but we had to forge on to Mount Vernon, Ohio before we could land and get out of our warm clothes. Once on the ground, we found that the temperature was 95 degrees--no wonder Ann was getting heat prostration.

After changing into summer clothes, we headed west with Ann in the back seat where it was breezier. We stopped again at the Lima and Rochester airports for fuel. Finally, we landed at Joliet just before 9:00 P.M.. The Cub was put in a hangar for the night. I started calling around for a place to stay. All the nearby motels were full and the only available room was in a hotel downtown. One of the hotel employees came out and picked us up. Ann and I were very hungry and went out of the hotel to find a place to eat. Much to our dismay, we could not find any place open. The hotel's dining room was open, but the meals were very expensive. Checking my wallet revealed that I only had enough money for the hotel room, tomorrow's gas and a little bit more (this was in the days before I had credit cards). We settled for a couple of candy bars apiece. Then, we made a phone call to tell JoAnn where we were.

The next morning(Sunday), Ann and I had breakfast at the hotel. Afterwards, the desk clerk was reluctant to have someone take us back to the airport. After a bit of pressure, he finally relented and drove us there. The day had dawned clear without a breath of wind. We loaded up the Cub and took off. After refueling at DeKalb, the Cub acted like a horse headed for the barn and seemed to know the way home. Near Stoughton, we saw a large peace sign mowed in the middle of a green alfalfa field and a bit later "uff da" appeared in another.

Approaching Madison, we could see Lake Mendota dotted with the white sails of a regatta. Instead of flying directly across the lake, we flew over our house and circled it until Jo Ann came out and waved to us. In a few minutes, we were back on the ground at Waunakee. The flying time from Gaithersburg to Madison was 13 hours and 20 minutes, which yielded a ground speed of 56 MPH.

Electrical System for the Cub

Although the intercom had solved a lot of problems, communication with control towers via the marginal ten channel Regency radio was still a dicey affair. In addition, the FAA was changing radio frequencies quite often as well as putting in control towers at many small airports. The radio's days clearly were numbered. The Piper Cub was designed as a trainer to be flown primarily in the vicinity of its home airport. Consequently, the engine did not have a generator and the cockpit was not designed for a radio or an electrical system. The Regency radio had become a liability on cross-country flights and needed to be replaced with a modern aircraft radio. To do so, an electrical system for the Cub would be needed. How to add a generator to the engine was pondered for a while. Finally, an advertizement appeared in *Trade-A-Plane* describing an alternator system for the A-65 engine. I quickly contacted the advertiser and he had a perfect solution. A parts kit and the Supplemental Type Certificate(STC) were purchased from him. A standard GM automotive alternator was fitted between the bottom of the crankcase and the top of the carburetor air intake. The alternator was belt driven by a pulley at the back of the prop. Now, I had more AMPS than horsepower.

A NARCO COM 10A aviation radio was purchased. Its radio tray was mounted vertically on two horizontal brackets attached to the rear of the front seat. The intercom was interfaced to the radio and it all worked perfectly. A piece of VELCRO was glued to the top horizontal radio bracket and corresponding pieces were attached to my plotter and a pencil. The strap-on knee board Joye had given me for Christmas completed my back seat navigator's station. Now, I had the tools of the trade within easy reach and everything was secure. I would no longer have to look for a map or pencil that had fallen on the floor. All I had to do was to learn how to use the radio. One of the deficiencies of my flight training was that I had received my pilot's license without ever having talked to a control tower.

Chapter5
The Prescott, Arizona Trip

Because of a lack of advanced courses at Prescott College, Ann was going to finish her chemistry course work at the University of Wisconsin. Thus, 1973 was to be her last year in Arizona. This provided me with a good excuse to fly the Cub to Prescott when the school year was over. Then, the two of us would fly the Cub back to Wisconsin. Ann, ever willing to participate in what promised to be an adventure, readily agreed to the plan. In order to shorten the initial part of the trip, I set a new route of flight from Madison to Grand Island, Nebraska. This route was Madison - Dubuque, Iowa - Marshalltown - Audubon - Fremont, Nebraska - Grand Island where the usual "southern" route would be followed to Denver. From there the route went south to Albuquerque, New Mexico and then west to Prescott, Arizona.

Route of Flight for the Prescott Trip

The Cub and I left Waunakee on Friday, May 25, 1973, at the usual "O dark thirty" on a cool overcast day. A down ski jacket and hiking boots were the uniform of the day. I had my personal belongings in a small backpack strapped in the front seat. The baggage compartment held the tie-down kit, a bag of tools, and window cleaning materials. Behind me on top of the baggage compartment were six items: my movie camera in its bag, the 35mm camera, the two man mountain tent, my sleeping bag, a vinyl pouch containing the maps, and a bag of candy and cookies. The weight of all this was far beyond the 20 pounds allowed for baggage in the Cub. The first gas stop was Dubuque, Iowa. The next several segments were routine except for Audubon where I was desperate for some lunch, but had to settle for the usual candy bar and a can of pop.

Flying across Iowa, I was surprised by the number of abandoned farmsteads. In many cases, the farm buildings had been razed leaving a small plot of rubble filled grass. In a few cases, a farm house desperately in need of paint was all that remained. In the past, a family could make a living on a 160 acre farm. Due to many years of very low prices for farm products that was no longer possible. Now a farmer requires several thousand acres to earn a viable living. Iowa was still the land of corn and hogs. Flying low, I could smell the hog pens as the Cub flew over them.

By the time I reached Fremont, Nebraska, it was late afternoon and the clouds had been left to the east. Some twenty miles past Fremont, there was a large column of smoke coming from a burning haystack. It reminded me of smoke I had seen on a daylight weather mission in Korea.

The T-6 "Mosquito" reconnaissance planes had found a large Chinese ammunition dump at the bottom of a very narrow mountain valley. They had called in several squadrons of F-84's to bomb it. Our B-26 route of flight took us along the front lines at 10,000 feet and the smoke was up to our altitude. As we approached the smoke, I could see a long line of F-84's circling the valley. One by one, they would peel off and dive down into the south end of the smoke filled valley. Somewhere in the smoke, they would release their bombs. In a few seconds, the F-84's would come out of the smoke, in a steep climb, at the north end of the valley. The commies were getting a good thrashing on that day.

Thinking back on the incident, I thought it would be good sport to make like an F-84 and fly through the smoke. I put the Cub in a shallow dive and entered the smoke. The sky quickly turned a dark grey. There were large blobs of red going past the Cub that looked just like 40mm anti-aircraft fire. When I realized the red blobs were large chunks of burning hay, I was jolted out of my enraptured state. In that the Cub was fabric covered and painted with highly inflammable dope, a chunk of burning hay lodged in any one of many places could set the airplane on fire. Any thoughts of mimicking an F-84 were tossed aside. The order of the day was to get out of the smoke. I threw the stick to the front left to initiate a diving turn and gave full throttle. In a few seconds, I was back in clear sky, leveled out and thinking a lot less of my idea.

By the time I landed to refuel at Grand Island, it was early evening and there were still 125 miles remaining to North Platte. The weather was very good and wind had not been a problem. The last leg should only take about two hours. Since everything seemed to be in order, I took off on the familiar route past the railroad roundhouse at Ravenna and out into the prairie. About an hour and a half later, light from the setting sun was blocked by a wall of clouds and it was getting dark rapidly. I had anticipated landing at North Platte just as the last light of day was fading. Now, it was getting dark much sooner than I had expected.

The Cub and I now were well beyond the point of no return to Grand Island. From the ferry flights, I knew there were airports at the small towns strung along the Platte river. With this in mind, I kept plugging along on the course line. Some 30 miles east of North Platte, it started to rain lightly, a solid line of thunderstorms lay across my path, and daylight was fading fast. It was obvious that the Cub and I were not going to reach North Platte. A quick check of the map showed that the Gothenburg, Nebraska airport was only 12 miles away, so I headed for it. By the time the Cub was a few miles from the town, daylight was gone, it was raining heavily with considerable lightning, and the wind had picked up to 30 MPH. Suddenly, the highway into Gothenburg appeared out of the gloom. By flying a 100feet above the highway, there was just enough visibility to keep track of where I was.

In that I had flown half of my missions over North Korea at night, my extremely good night vision was a real asset. In the starlight, I could readily identify rivers, mountain ridges, and other landmarks. The Chinese on the ground were very careful not to let any light show. If they did, a B-26 would come down and put it out. On many nights, there would be a cloud deck above us cutting out the starlight and it would seem as if the B-26 was flying in a black pool. While sitting in the clear nose cone, I would get vertigo. I would ask the pilot to fly level, which of course he was. I would have to climb back through the tunnel and into the right hand seat. The artificial horizon was watched until up was up and then I went back into the nose. After a while my body would again tell me that we were in a 45-degree bank and the process would be repeated.

On one night mission, we encountered a "Lamplighter" and a squadron of Marine F4U's attacking a truck convoy. The "Lamplighter" was either a C-47 or a PB4Y-2 circling at 10,000 feet and dropping parachute flares. The flares burned for three to four minutes while lighting up the countryside like it was daytime. During this time, the F4U's would dive down to strafe and bomb a truck convoy. Since the flares would go out suddenly, a pilot on a run would be flying in daylight one second and an instant later would be in night again. Every time I watched an F4U start its run; I hoped, for the pilot's sake, that the flare would stay lit.

Tonight, the lightning flashes were my equivalent of the flares. When one flashed, I could get a quick glance at the instruments. If I was looking down, get a peek at the ground. By following the highway to the southeast, I finally reached Gothenburg. In the pouring down rain, the village lights were a welcome sight as was its name on the silver water tower. At the southeast side of town, the airport's runway lights were on. As I flew over the airport, I made a quick confirmation of the wind direction and made a tight 180 degree descending turn. Just as I leveled out, I chopped the power and flared to a good landing. I taxied in, found a tie-down spot, and climbed out of the Cub. As I was tying it down, a man in a raincoat appeared out of the dark to help me. He said that he was quite surprised that I could taxi a Cub in that much wind. By the time I landed, I had been en route 13 ½ hours. In a Cub, that seems like a lifetime.

Once inside the building, I was introduced to two fellows, who had landed a few minutes earlier in a Piper Tri-Pacer. In all the excitement, they had forgotten to close their flight plan to Holyoke, Colorado. One fellow's wife, who had not heard from the pilots, had called the Akron FSS. The briefer had started to call airports along their route. The Tri-Pacer pilot was surprised when the FSS briefer located him. A quick call to his wife settled things down for her.

The airport manager gave me the keys to a beat up old car. The three of us went to an equally beat up motel on the edge of town. I called JoAnn, told her where I was and related the events of the day. Fortunately, the motel had a small restaurant. The three of us were able to talk the cook into making us something to eat even though it was long past closing time.

The next morning(Saturday), the three of us had a good breakfast then drove the mile or so back to the airport. Due to a low cloud deck the weather was not the best, but there was reasonably good visibility underneath it. I decided that it was good Cub weather. The Tri-Pacer fellows were going to wait until the ceiling was higher. I plotted a route of flight directly from Gothenburg to Julesburg--109 miles away.

After taking off, I flew the Cub across the Platte River and approached the south river bluffs. Right on the course line, there was a small reservoir just beyond the bluffs. A small notch in the bluffs provided some better cloud clearance, so I headed through it. As I reached the water, I got a surprise. The hills on the other side of the reservoir were in the clouds. Now, I was in a real predicament. In front of me the way was blocked, there was not enough room to turn around, and I had about thirty seconds of flying time before the far hills were reached. Looking to the right, I saw the earthen dam that held back the water. There was only about 50feet between the dam and the cloud deck. My choices were becoming very limited. Making a flat turn, I headed for the dam. The Cub went over it in a shallow dive and out into the valley of the Platte River.

Once in the valley, there was better cloud clearance, so I headed for Lee Bird Field at North Platte. By the time the 20 miles to North Platte had been flown, the ceiling had gone up to a 1,000feet. After landing and topping off the tank, I checked the weather at the FSS. The maps did not show any thunderstorms. However, the briefer did say that it was starting to get windy with 35 mph being reported at Sidney, Nebraska. Not having the faintest idea where Sidney was, I let the comment go on by me. Had I checked my map, it would have shown that Sidney was only 25 miles north of my course. The first part of the route to Julesburg followed Interstate 80. It quickly became obvious that a 45-degree wind correction angle was necessary to stay on course. This should have told me something, but in my mind it registered just as another windy day.

As I watched the cars on the interstate go past me at a good clip, it was clear that the Cub's ground speed was quite slow. I was so busy keeping the Cub on course that I did not even think about computing what it actually was, but it wasn't much. Since Julesburg was still some 80 miles away, having enough fuel to get there was added to my list of concerns. Refueling the Cub at North Platte was a good idea as flying non-stop from Gothenburg to Julesburg was out of the question.

As the Cub slowly worked its way along the course line, long plumes of blowing dust were dancing across the farm fields. By the time the Cub reached Julesburg, the wind was really howling. I made an approach to a runway that fortunately pointed right into the wind. As I started to flare, the wind slowed me up so much that I felt the Cub would not stay on the ground. Pushing full throttle, a go-around was initiated. On the second approach, the landing did not feel any better, so it also was aborted. If I could not get down on the third try, I would fly straight downwind to the airport at Holyoke, Colorado that was only 27 miles away. Hopefully, the high winds would not have reached there yet.

On my third downwind leg, I saw a group of men run out of the International Harvester building, jump into a pickup truck, and head for the runway. They were going to help me, so I made another approach. This time I flew the Cub down to about two feet off the ground and slowly worked my way up to the wind tee alongside the runway. At this point, I started to fly formation with the wind tee. As it changed heading, so did I. One of the men went to a large hangar directly in front of me and opened the doors. Two men went under each of the wings and two to the tail. When they all were in position, they reached up and literally pulled the Cub out of the air. With the six men hanging onto the Cub, I taxied ahead and went directly into the hangar.

Once inside, I got their side of the story. The business owner was a fellow, named Perry Case, who flew a Mooney 21. He heard me fly over and after the second pass knew that I was in trouble. Quickly rounding up his coworkers, they headed for the runway. I told him that I had planned to make one more pass and then head for Holyoke. I also told him how grateful I was for their help. He said I would be here for a few days as these kinds of winds don't die easily. Then, he handed me the keys to his brand new Ford pick-up truck. He said to use it until I was ready to leave, just put the keys under the floor boards. After thanking him, I took the men back to the International Harvester store and drove to a motel along the Interstate.

I called John, in Denver, relating another Julesburg adventure. Checking my maps, I found that it took three hours of flying time to go the 115 miles from Gothenburg to North Platte and then to Julesburg. That's a ground speed of a paltry 38 MPH. The next day the wind was still howling. The local paper told of semi-trucks being tipped over on Interstate 76. I drove out to the airport to check wind conditions and the Cub, but there was no way I could leave. John was scheduled to go to Sweden in a few days to visit our sister. If I had to stay here much longer I would miss him in Denver.

Finally, on the third day(Monday) the wind had disappeared and it was warm and sunny. I parked Perry's pick-up alongside the hangar, put the key under the floor mat, rolled the Cub out, refueled it, and departed for Sterling. After an hour flight, I landed there.

After Sterling, the ground gains altitude quite rapidly. Denver's JEFFCO airport sits on top of a plateau at 5,654feet. Approaching JEFFCO, I called the tower on the radio and received no reply. As I got closer, I called again and there still was no reply. Finally, I entered the traffic pattern and was given a green light to land. When I was tying down the Cub, the line boy came up to me and said that the tower wanted to talk to me. Inside the FBO's office, I called them on the telephone. I told them that I had called them numerous times, but could not raise them. The control tower operator asked me what frequencies I had used and I related them. None of them were currently in use, then he asked me where I came from and I told him Julesberg. At that, he decided I was a country bumpkin not used to the big city. He said to be better prepared the next time.

When John got to the airport, we checked the radio system and discovered that the battery was dead. After several hours of trouble shooting the system, we determined that the voltage regulator was burned out. Because of this, the battery had run down during the course of the trip. John and I drove all over Denver trying to buy a new regulator, but because it was Memorial Day no auto parts stores were open. This meant that when I departed the radio would be inoperative. To circumvent this problem, the next morning John took me to the airport before the tower went into operation. After loading my bags into the Cub, I taxied to the runway and took off.

The route of flight from JEFFCO to Pueblo, Colorado was to the south along the front range of the Rocky Mountains. The only big city along the route was Colorado Springs. Staying to the west of Interstate 25, I avoided both the city and the near by military installations. While doing so, I flew past the front entrance of the U.S. Air Force Academy whose buildings could be seen in the foothills of the Rocky Mountains. Before I left John's house, I had called the Pueblo control tower to tell them I had no radio and the time I would arrive. After an hour and 55 minutes, I entered their traffic pattern, and the tower gave me a green light to land.

After tying down the Cub, I took a taxi into town to the Chevrolet dealer and bought a new voltage regulator. On long cross-country flights I always carry a canvas bank bag full of small tools for making minor repairs. Using these, I was able to install the new regulator. Although working on both sides of the firewall meant lots of trips into the cockpit to re-tape wrenches to the bolt heads. When finally finished, I was so engrossed in communicating with the tower that I almost taxied into a parked fuel truck. I had to climb out and pull the Cub back in order to taxi past the truck. It would have been very embarrassing to have turned the prop into kindling wood on the side of the truck.

Departing Pueblo, I had planned to fly sixty miles south and then go west over the 9,500foot La Veta Pass to Alamosa, Colorado. This would put me behind the Sangre de Christo Mountains. Once there, I could follow a large valley all the way to Albuquerque, New Mexico. As the Cub approached the point where I would have to turn west to the pass, it became apparent that the pass was in the clouds. I would have to change my plans. The next town was Raton, New Mexico on the south side of the 7,843foot Raton Pass. Fortunately, Interstate 25 went directly there and all I had to do was follow it. I was getting into country higher than I had ever flown over. The good old Cub just kept put-putting along and slowly gained the needed altitude. On the east side of Raton Pass there was a table-topped mountain peak that was several thousand feet higher than the pass. As the Cub went over the pass, I looked up at the peak and the series of rock ledges that formed a staircase leading to its top.

The Raton airport was next to a large grandstand at a horse race track. After landing, I called the fuel number and was told that someone would be along shortly. After twenty minutes, I called again and got the same reply. After an hour and a half, an old man finally appeared and put fuel in the Cub. The delay meant I would not reach Prescott today.

The next stop was Las Vegas, New Mexico--90 miles away. The airport was near Interstate 25, so all I had to do was follow the highway and enjoy the scenery. The color of the countryside was the typical high desert light beige with large areas of grayish green bushes scattered about. Other than the Interstate, there were few signs of civilization. Some forty miles before Las Vegas, a fine mist of oil started to appear on the windshield and on the left wing struts. Although, I was over the Interstate, it was no place to be running out of oil or attempting engine repairs in some field. The oil did not appear to be hemorrhaging badly, so I decided to keep on going to Las Vegas.

Once on the ramp, I checked the engine and found several of the screw's holding the #4 cylinder rocker cover were loose. A couple of turns with a screwdriver cured the problem. After checking the oil level, cleaning the windshield and wing struts, as well as refueling the Cub, I was ready to go again.

After taking off, I followed Interstate 25 some 20 miles southwest to where it went around the southern end of the Sangre De Christo Mountains. From there, the Interstate disappeared from sight as it went towards Sante Fe, New Mexico. To the east of Albuquerque, New Mexico there is a narrow mountain range that is 60 miles long. The northern end of this range is 15 miles south of Sante Fe. I had flown along the face of the Rocky Mountains several times, but the immense scale of what now appeared before me was overwhelming. The two mountain ranges loomed high to the right and to the left and both slanted down to an expanse of desert that is at about 7,000feet above sea level. The gap between the ends of these two mountain ranges provides a path through the mountains.

Where the Interstate and I parted company, the Cub was pointed straight west to go across the mesas and through the gap. The first thing to greet me was the east face of a tall mesa. To start climbing, I gave the Cub full throttle, but did not gain much altitude. I kept looking at that face and wondering if I would ever see over it. Finally, I could see beyond it and flew over the cliff face with a hundred feet to spare. As I passed over the town of Galisteo, New Mexico, I could see a large hangar at the Santa Fe airport some 20 miles away. In another few minutes, I linked up with Interstate 25 again and followed it south to Albuquerque's Coronado airport that was at 5,270feet.

During my Navy V-5 days, I had attended the University of New Mexico in Albuquerque. At that time there was a small airport located east of campus that had a dozen war surplus Piper L-4's. On weekends, I would rent an L-4 for an hour. One day I took an L-4 up to 10,500 feet and decided to spin it down. I pulled the nose up and before I could kick the rudder the Cub started to tail-slide. I knew that Cubs had a reputation for shedding their elevators in a tail-slide. As I plummeted down backwards, I hoped that the elevators would not part company. In the thin air, the L-4 must have dropped several hundred feet before the tail popped up, the nose pointed down, and I had some semblance of control. Little did I know that twenty seven years later I would again land a Cub at Albuquerque.

In those early days, there also was a small airport on the west side of the Rio Grande River. When I could get a ride over there, I would rent a plane. They had a strange collection of airplanes. One day I rented a Porterfield that I had never flown before. The instructor signed me off after only 15 minutes in it. Keeping in mind that at the time I had only 18 hours of flying time and a Student Permit. Later in my Air Force navigator cadet days, we would fly night celestial navigation missions from Houston, Texas to Albuquerque's Kirtland AFB.

The next stop on the route was Grants, New Mexico. The flight path was over desert country with mountains off in the distance and mesa's alongside the route. To the north, the ground sloped up to the base of the mesas whose faces were cliffs with horizontal strata that displayed various shades of light brown. The individual mesas were quite tall and separated from each other by small canyons. I took a very nice photo of these mesas that now hangs in my office. Although I was still wearing my ski jacket while flying, it was near 85 degrees on the ground.

After landing at Grants, altitude 6,520feet, I checked the weather at the FSS. The briefer said the density altitude was 9,000feet which gave me pause as I had never taken off at such a high altitude. He told me that as I took off to the west go over the golf course as it usually provided some lift. I followed his advice. After a few minutes of mushing around at 50feet, found lift and the Cub gained enough altitude to proceed on course. By now it was getting quite late in the day, but it was only 54 air miles to Gallup. The Cub was climbed up to 8,000feet as we proceeded toward Gallup.

At 30 miles out, I called in on 122.8, as Piper 88354, and asked for an airport advisory, which was duly received. Finally reaching Gallup, I made a straight in approach, touched down at 7:15 P.M., and taxied the Cub to the ramp where a fuel truck was waiting. When I went in to pay for the fuel, the man at the desk said that the office usually closed at 6:30 P.M.. But my radio call was so clear and strong they thought I was a twin engine Piper Navajo. They had waited to make a large fuel sale. Needless to say, my three and a half gallons was not what he had anticipated.

I took my backpack and the radio out of the Cub and proceeded to walk the couple of city blocks to the nearest motel. The sun was just setting and its rays were reflecting off the high cirrus layers some of which turned red and gold while others were light grey against the dark blue sky. Walking into town, I realized there were very few non-Indians in sight. I felt just like I was in a western movie with Navajos everywhere.

Before I left home, a number of people had told me that, in the desert country, only a few early morning hours were good for flying. By midmorning it would be so hot and bumpy that flying would not be possible. Consequently, I was up bright and early the next morning(Wednesday) to greet another perfectly clear day with cool temperatures. It was 85 miles to Winslow, Arizona and again there was an Interstate highway to follow. After takeoff, I climbed up to 1,000 feet AGL.

Some 25 miles before Winslow, there was a coal-fired power plant in the desert whose plume of smoke went up, bent over, and then formed a 20 mile long trail to the northwest. Fortunately, the smoke plume was rather narrow. Non-the-less, it despoiled the beauty of the desert. Winslow was in a valley with the airport on the southwest side of town. While the lineboy put fuel in the Cub, I looked at a very nicely restored white airplane of a type I had not seen before and could not identify. I took a picture of it and later determined it was a Commonwealth, actually a postwar Rearwin.

From Winslow to Prescott, I could not follow the Interstate any more as it had veered northwest toward Flagstaff, Arizona. Off in the distance, I could see the snow capped 12,633foot Humphrey's Peak in the San Francisco Mountains. Fifteen miles west of Winslow was the famous Meteor Crater National Park. I deviated a few miles off the course line to take a look at the crater. I was surprised at how large and deep it was. There was a trail leading down the side of the crater to a small building at the bottom, presumably an information center. After inspecting the meteor crater, I worked the Cub back onto the course line and started to climb up to 9,500feet to clear the 8,000foot mountains to the west. The next big landmark was Mormon Lake. The map showed it to be a dry lake, but it turned out to be full of water with lake cottages along the northwest shore.

Then, the Cub and I entered some of the most beautiful desert country in Arizona. The land became a series of tall rock spires and ridges composed of multi-hued horizontal strata. The colors of the strata ranged from sand to bright red. There was no pattern to the spires nor was there a place to land the Cub, either on them or between them. Off to the right, was the famous Oak Creek Canyon that draws visitors from all over the world. The canyon worked its way north through the mesas and its walls had considerable tree cover.

The Sedona airport resembled an aircraft carrier as it occupied the complete top of a large mesa. The mesa itself had rather steep slopes and far below on the left side was a narrow river. The small town of Sedona was at the base of the right side of the mesa. I was perched up in the sky at 9,500feet and the scenery was so beautiful I felt compelled to tell someone about it. I called up the Sedona airport on 122.8 and proceeded to tell the voice on the other end how beautiful it was up here. He talked to me for a few minutes, but did not seem to appreciate what I was experiencing.

Just past Sedona, the route of flight went over a twenty-mile wide valley with the town of Cottonwood alongside the Verde River. To the west, there was a mountain with an abandoned open pit copper mine on its upper face. Alongside the mine was the former ghost town of Jerome that was making a comeback as a tourist attraction.

As I flew past Jerome, I could see the airport at Prescott and off in the distance Granite Mountain defined the horizon. Since I was still at 9,500feet and the airport was at only 5,042feet, a gentle let down was initiated. After being so high in the sky for so long, I made a bad approach to the runway and had to make a go-around. When I pushed the throttle forward to get enough power to make a missed approach, the Cub started to behave very sluggishly. It staggered along at twenty feet above the sagebrush and I could not get it to gain altitude. My first reaction was that I had let down into some hot air at the desert floor and the density altitude was the cause. After a couple of flat turns, the Cub was lined up with the runway again and I landed reasonably well. As I taxied up to the gas pumps, I noticed that the carburetor heat was still pulled out--the sluggish climb mystery was solved.

The 1,643 miles from Madison to Prescott took 33 hours of flying time. The overall ground speed was only 50 MPH. Ann and one of her friends drove out and picked me up. We stayed in an off-campus house where a number of Ann's friends were living.

The Return Trip From Prescott to Madison

I knew that Ann and I were going to fly over some very high country on the way home, so we got rid of everything except absolute essentials. We kept: a minimum of clothing, the tent's rain fly and poles, our sleeping bags, the cameras, and a gallon of water. The surplus items were put in a box that her pals would ship home after we were gone. I had the luxury of sleeping in one of the few beds in the house. Ann and I arose the next morning at 4:00 A.M., ate breakfast, and then went to the airport before sunrise.

It was Thursday, May 31,1973, when we look off into another perfectly clear day. I was in the front seat as pilot and Ann was in the back doing the navigating. The route of flight to the Grand Canyon Airport was a few degrees to the right of north and was over high desert. So we could confer about where we were, both the pilot and navigator had maps. Forty miles out, we flew past a 9,256foot mountain that was just to the right of course. Once beyond this point, the navigation check points started to get rather scarce. Fortunately, a railroad wandering through the desert crossed our path about 2/3rds of the way to the Grand Canyon. Due to Ann's navigational skills the Cub flew over the tracks right on time.

The Grand Canyon airport was very picturesque having a rustic looking operation's building with tall pine trees around it. The Cub was landed on runway 18 into a 20 mph wind. I asked the fellow in the airport office about flying over the canyon. He said it was no problem and we could fly down in it if we so desired. We refueled the Cub, climbed in, taxied to the north end of the runway, and started our takeoff. After rolling a long way, the Cub was finally airborne at 15feet off the ground, and then it sailed through a gap in the trees. Unfortunately, that was all the higher the Cub felt like going. Luckily, the ground sloped downhill to the south.

Now, Ann and I were flying only a few feet above the tree tops and needed to gain some altitude. The

Cub can be coaxed into climbing by giving full throttle, initiating a shallow turn with the nose a few inches above the horizon. With a lot of such coaxing, the Cub finally achieved a few feet per minute rate-of-climb. After some ten minutes of anxiety, a safe altitude was reached.

In order to do some sightseeing, I had wanted to fly east along the south rim of the Grand Canyon. The strong south wind was such that if we flew too close to the edge there was a possibility of the downdraft pulling the Cub into the canyon. Once inside, we would have to follow it to the northeast end before getting out. That was not an appealing prospect, so Ann and I flew the Cub a half mile from the rim and over solid woods. We could see the north rim of the canyon, but not the bottom. Because of the abundance of landmarks and the unlimited visibility, all that Ann needed to do was give me a compass heading. Once pointed the right direction, I could see the next checkpoint even though it might be 20 miles ahead of us.

Twenty-five miles along the route we flew over the canyon of the Little Colorado River. It cut through flat country, was some 3,000feet deep, about 200 yards wide at the top, and the greyish brown canyon walls were vertical. In the dark at the very bottom of the canyon, was a small river. The canyon zigged and zagged back and forth very tightly as it headed toward the Grand Canyon.

In another fifteen minutes, we came to the south end of the Echo Cliffs. The Cub was turned north to follow them to the town of Page, Arizona. Now, the Cub was flying over the Painted Desert that lived up to its name. Beneath us it looked as if someone had painted streaks on the desert floor with long strokes of a giant paint brush using many different shades of red, brown and yellow.

The Echo Cliffs, to our right were formed by an uplifting of the ground The resulting ridge was slanted on the east side, but the west side was a tall cliff. At one time, there had been a mine of some kind cut along a section of the west face of the ridge. The cut revealed a horizontal rainbow of bright colors: blues, turquoise, reds, yellows, and purples. As Ann and I flew north along these cliffs, the Colorado River could be seen to our left.

In her third year at Prescott College, Ann had gone on a three-week rafting trip through the full length of the Grand Canyon. Ann pointed out the place where she had jumped overboard to ride through a small rapids in her life vest. At a later point in that trip she was washed overboard in a large violent rapids. She came up under the raft, one of the persons on the raft saw her blond hair in the water, grabbed it, pulled her out from under the raft and then onto the raft. Later, Ann said that even though she was trapped under the raft, she decided it was not her time to die and would get out of her predicament.

To the west of the Colorado River is a 20 mile long cliff that faces east. From the river, the ground slopes up to a lower set of cliffs. At the top of these cliffs, the ground slopes up again to the base of much taller upper cliffs. Both levels of these jagged cliffs had horizontal stripes of dark red rock. Hence, the name: Vermillion Cliffs.

At the north end of the Echo Cliffs, we would have to fly over a saddle in them to reach Page, Arizona. Since the Cub was flying lower than the top of the cliffs, a slow climb was initiated to get enough altitude to clear the saddle. Finally, the town of Page came into view. Then, Ann and I could see the Glen Canyon Dam that held back the Colorado River forming Lake Powell. The town had been built as a temporary village to house the construction workers. When the dam was completed, the town became a permanent entity. We landed easily. Since Ann and I were going to camp overnight, the Cub was tied down on the ramp. I can not recall why we only flew two legs this day as that was atypical for me.

It was a beautiful sunny day with a few scattered cumulus clouds, so we decided to walk the mile to the dam. By the time we got there, both Ann and I had parched throats and were quite dehydrated. The 85 degree heat and the low humidity had taken a much larger toll than anticipated. I was glad that a gallon of water was carried in the Cub. This short walk convinced me that a gallon was probably not enough, but we could not afford the extra weight. The dam was 360feet tall and on the upstream side the water was 285feet deep. Ann and I rode the dam's elevator to the lowest level. We went outside into the downstream canyon where the air was quite cool. The canyon's walls were a deep vermillion color with some nearly black sections. A couple of 28 foot long Army surplus rafts "pig boats" were tied up along the riverside waiting for a commercial raft trip through the Grand Canyon.

After examining the insides of the dam, we walked across the top of it and witnessed an unusual event. Ann and I were looking at the water when a large Crow came diving down from the south. As he reached the upstream side of the dam, he partially folded his wings, gave a loud "caw," did a perfect slow roll, and pulled up into level flight. It was obvious that the Crow loved to fly as much as we Piper Cub pilots do.

After supper, Ann and I went to the north end of the runway, walked fifty yards down the hillside to a level spot, set up the rain fly as a tent, and rolled out our sleeping bags. During the night, a thunderstorm arrived with heavy rain and strong winds. Since the front of the tent was open to the wind, wet sand started to blow in on us. I pulled the front closed and as soon as I did, Ann started to get claustrophobic and insisted that I open it. Each time I did, the wet sand blew in and when I closed it, Ann made a fuss. Finally, I gave up, pulled my sleeping bag over my head and went to sleep.

The next morning(Friday) the storm was past and every thing in the tent had a layer of wet sand over it. I can still feel the grit in my teeth and the sand my hair. It took us quite a while to clean things up, and get the gear packed in the Cub. Then, we checked the weather via the telephone with an FSS in Utah. They indicated a front was coming through with low ceilings and rain within an hour or two. Deciding to play it safe, Ann and I waited a couple of hours, but the weather stayed about the same. It was rather dark off to the northwest with low clouds, but there was high cloud cover and unlimited visibility on the south side of the lake.

Finally, Ann and I decided just to go on a local flight to see the sights along the Lake Powell shoreline. Once in the air, the weather along our intended route of flight looked very nice. I called the Page Unicom, on 122.8, and had them file a flight plan to Blandling, Utah for us. I had planned a gas stop at Oljato, Utah, but forgot to mention it when I filed the flight plan. Before we took off, I had talked to an oil company pilot who flew a Citabria in these parts. He told me that the Oljato airport was down in a small canyon and rather hard to find if you were not familiar with the countryside. Keeping this in mind, we proceeded along our flight path toward Blandling.

Ann and I were flying along the north side of the 10,388foot Navajo Mountain. There was a solid mass of jumbled lava flows that went from its peak all the way to the lake. This was some of the wildest terrain I had ever seen. The beige colored rock was perfectly smooth in appearance. In the midst of jumble of lava were several red colored spires and ridges. One of the interesting sights was Rainbow Bridge, a large high arch of stone, carved out of the rock by erosion. Due to the dam, the water was slowly creeping up toward the bridge. There was some concern that when the water reached it, the bridge would be destroyed.

As part of her freshman orientation week at Prescott College, Ann had been dropped off on this side of Navajo Mountain to camp alone for three days. She said that it was a wonderful experience, but after three days of complete solitude, she decided that she needed to be with people. While flying over the seemingly endless lava flows, Ann sighted the spot where she had camped and excitedly showed it to me.

Past the lava flows, the country turned into a series of tall flat topped ridges, separated by deep canyons, that went out of the high country to the lake. By now Ann and I had been flying an hour, were some 50 miles from Page, and I had only a vague idea as to where we were. I started looking for the Oljato airport. We flew along one of the ridges until it ended in a box canyon, turning back toward the lake I went over the next ridge and repeated the search. All of this was fruitless and both of us were getting concerned. Enough fuel had been used searching the ridges that we were beyond the point of no return to Page. The next airport with fuel was at Blandling some 59 miles away and was out of our reach. I knew where we were in general terms, but could not pinpoint our location relative to the airport. Off toward Monument Valley I could see a paved road, but my map only showed a dirt road in this area. The map also showed a dirt airstrip to the east, but it did not have gas and was not attended.

The Cub had been in the air an hour and 45 minutes and was getting low on fuel. When I saw a set of tire tracks on the desert floor, I decided it was better to be on the ground with some fuel than in the air without any. I told Ann we were going to land. She had never made an off-airport landing and was not too thrilled about the idea. I made one low pass to check for how firm the dirt in the tracks appeared. They looked solid, so the Cub was landed on the tracks. Once on the ground, two Navajo women came walking over a ridge above the road and came down to us. I told them we were looking for the Oljato airport, while badly mispronouncing its name. One of them simply said "down the wash" and they walked off before I could get clarification.

Some information was better than none, so Ann climbed back in the Cub, I propped the engine, climbed in, and we proceeded to take off. Just as the Cub broke ground, I realized that the tracks went up a six-foot embankment. The Cub's wheels cleared it only by a foot or so. Once in the air, I flew in the general direction that the Indian woman had pointed. In a few minutes, we were over a large green depression in the ground that made a good landmark. In desperation, I turned the radio to 122.8 and gave a call to the Oljato airport. Much to my amazement, the reply was loud and clear. I explained where we were, the voice told us to fly east, and turn left around the edge of a low mesa. Following the directions brought us to the airport. It was only a minute's flying time away from where we had been. Ann and I were greatly relieved to get back on the ground and be able to get gasoline.

The only building at the airport was an Indian trading post built in 1909. To go inside was to step back in time. The ceiling was low and its beams were 18 inch diameter logs. The room was dimly lit, the walls were smoke stained and trade goods were stacked everywhere. The owner was a white man, with a Navajo wife. He didn't get many fly in visitors and was delighted to see us. He told us that the green depression was the remains of an underground uranium mine. The depression had been the last pillar in the mine. When they mined it, this section of the roof of the mine fell creating the green depression. He referred to it as the million-dollar hole as that was the value of the uranium in the pillar. As much as we enjoyed visiting with the man and his wife, I knew we were overdue on our flight plan. He had no telephone, so Ann and I needed to get to Blandling before the FSS started to look for us.

Just after taking off, we flew the Cub alongside a mesa and then went out into Monument Valley. It is one of the most spectacular sights in the world. The valley contains a large number of isolated mesas capped by rock spires that rise five hundred to a thousand feet above the valley floor. They are widely spread and stand out against the sky, especially from our vantage point only a few hundred feet off the ground. To the east, the ground rose quite a bit and there was a paved road going to the high point that was capped by tall rock formations on both sides of the road. Since our flight path was to the east, we started to follow the road. From the back seat, Ann started to sing the song Follow the Yellow Brick Road from the movie the Wizard of Oz and I quickly joined in on the verse. The song was really appropriate as the countryside was like another planet. We had little idea of what lay ahead on the road. Perhaps the Wizard did live on the top of a mesa. In addition, it had started to rain lightly, which added to the mysterious atmosphere. Monument Valley was the most awe inspiring sight I have ever seen from the Cub.

Several years later, I was riding the airlines from San Diego to Chicago and the route of flight was just south of the Grand Canyon. When I looked out of the window, I could see Navajo Mountain, the road that Ann and I had landed on, the million dollar hole, as well as the Oljato airport. From 30,00 feet, it was clear why we had such a difficult time finding the airport as it was surrounded on three sides by mesas.

As we left Monument Valley, there was a small town called Mexican Hat. The edge of the ridge behind it contained a red, up and down, zigzag stripe that looked just like those on a Mexican sombrero. In college, Ann had visited the town and was very excited to see it from the air. The highway underneath us twisted and turned through a number of valleys and went over several mesas, but we could generally follow it to Blandling. The flying time from Oljato to Blandling was an hour and five minutes. After landing, we parked the Cub and Ann tied it down. The sky to the west was filling with some rather black cumulus clouds whose tops could be seen in the distance. We were not going any further today.

The FBO came over and told us that the FSS had started to call airports looking for us. He was amazed that we were flying a 65hp Cub. He had not seen one in this part of the world for some twenty-five years. I went in the office and called the Utah FSS to inform them that we had reached our destination safely.

Even though my radio was working quite well, I always have trouble raising FSS stations, as a result it never entered my mind to give them a call en route. The nearest FSS was at Farmington, New Mexico. There were several mountains between us and them. I'm sure they would have never heard me and then I would have become uncomfortable at not being able to make contact. As a result, the radio was rarely turned on during the cross-desert flight segments.

The lineboy's father ran a motel in Blandling. When the work day was over the lineboy drove us into town. The motel was just slightly worse than the one in Gothenburg, Nebraska. It was a long low 1940's wood building with 15 units all in need of paint, but the rate was only $12, so we couldn't complain. It is worth mentioning that at Blandling we could still see Humphrey's Peak. For three days, I had been using this mountain as a point of reference even though most of the time it was more than a hundred miles away.

The next morning(Saturday) the sky was overcast, it was quite cool, but the visibility was reasonably good. Ann and I loaded up the Cub and took off to the north where a solitary pair of mountains, Mount Mnnaeus and Abajo Peak, towered over the countryside. Their tops were roughly 5,000feet above the surrounding ground. As we started past the east side of these mountains, the visibility decreased rapidly, and rain started to fall. The front that the weather forecasters had told us about the day before had decided to move. The town of Monticello, Utah was just ahead and five miles beyond it was the San Juan County airport. When the airport came into sight, I made a long straight in approach and landed. After tying down the Cub, Ann and I went into the unlocked airport building only to find that nobody was there. We settled down to wait out the weather by reading the usual old magazines in the pilot's lounge. These magazines are even older and more beat up than those in a doctor's office.

In a half hour, the fellow, who ran the Blandling airport, showed up. He figured that we had not gotten very far and had driven up to see if we had reached San Juan. It was nice of him to have been concerned about our fate. Ann and I learned that he ran both airports and made his living flying tourists over the Canyonlands National Park and Monument Valley in a Cessna 185. Around 11:00 A.M. the clouds broke up. Within a half hour it was a bright sunny day with only occasional fair weather cumulus clouds dotting the brilliant blue sky.

Ann and I fired up the Cub again, took off, and headed north for the town of Moab, Utah only 65 miles away. To our left was a rather high north-south ridge separating us from the Colorado River and the Canyonlands National Park. Since the airport at Monticello was at 6,996feet, after takeoff we had enough altitude to see over the ridge and out into the Canyonlands. Fortunately, all Ann and I had to do was follow a highway and enjoy the scenery. To the east, we could see the snow capped tops of 14,000foot mountains. The general slant of the land in that direction was upwards. As far as I could see to the west, there were mesas, canyons, and desert. At Moab, we crossed the Colorado River that was above flood stage. It had covered the low-lying land with a sheet of dark brown muddy water. The river went through a cut in the ridge and into the Canyonlands where it turned south and disappeared from view.

The Moab airport was fifteen miles northwest of town, had a long runway, and was situated in open country. From the airport, we could see the high ground on either side of the broad valley of the Colorado River as it wended its way west from Grand Junction, Colorado.

After taking off from Moab, we flew to the northwest to avoid a high mesa just to the east of the airport. This direction was flown until a railroad track was sighted and then we followed it to the northeast. After an hour of following the railroad, a river bend was rounded at the town of Mack, Colorado and the Cub entered the fruit growing area west of Grand Junction. The visibility was unlimited and on each side tall mountains could be seen off in the distance.

The Grand Junction airport was to the northeast of the city, sat on top of one low mesa and below a higher one. Unfortunately, the airport had a control tower, so we had to go through the drill of calling in before landing. Grand Junction was a fairly busy stop, as United, Western, Frontier, and Continental airlines all served the city.

Ann and I refueled the Cub, had a quick lunch in the airline terminal, and then climbed back into the Cub. As we taxied out, a Western Airlines B-737 jet pulled out and followed us down the taxiway. When the tower operator gave us clearance to takeoff he also said to turn left as soon as we could. By now it was getting hot, the airport was at 4,857feet and as the Cub broke ground, it gave signs of not wanting to climb. To the left were a number of hangars and beyond them were the outskirts of town. Since there was a bit of a crosswind from the left, I slid over toward the embankment on the right side of the runway to see if I could find some slope lift. About the same time, the tower called telling me to turn left. In that I was busy trying to get some altitude, I ignored him. That brought another rather annoyed message to turn left. This time I told him I would turn when I could, but that we were barely airborne. Finally, the Cub reached an altitude of about 40feet and I could initiate a turn to the left. Just as the Cub cleared the runway, the Western 737 came whistling past right behind us. I was very unhappy with the tower operator, whose only interest seemed to be in getting the Western flight off on time.

Our next destination was Eagle, Colorado in the heart of the Rocky Mountains. As the Cub proceeded northeast up the Colorado River valley, there was a very high mesa off to the north that had a nearly vertical face. To the south were some 9,000foot mountains with green grass and trees all the way to their tops. Fifty miles east of Grand Junction, we flew past the village of Rifle where there were two small airports alongside the highway. To the northwest of Rifle was a gigantic hogback stretching off to the northern horizon.

As the Cub approached Glenwood Springs, the Colorado River valley was getting narrower, and the hills on each side of it were getting higher. Once over the city, Ann and I could see the Aspen and Snowmass ski areas at the south end of the Crystal River valley. Our family had skied Aspen many times and often stayed in Glenwood Springs.

At Glenwood Springs, the Colorado, River comes out of a very narrow, fifteen mile long east-west canyon that has a railroad and a highway at its bottom. Since the canyon had a number of turns in it and a power line going back and forth across the river, it was not a good idea to fly through at a low altitude. At the same time, the tops of the canyon walls were at 8,000feet. After talking it over, Ann and I decided the best course of action was to fly east along the top of the south side as close to the canyon wall as we dared. This placed us in an interesting position. The right wing tip was just above the trees on the top of the canyon wall and below the left wing tip it was 2,000feet straight down to the river. Ann took movies out of both sides of the plane to illustrate the contrast. She shot some film out of the left side in which a school bus appears as a small yellow dot on the highway below. In 15 minutes, we were to the east of the canyon and the valley widened. The airport at Eagle lay before us. The Cub was landed back toward the west.

This was an interesting airport as it was the nearest one to the Vail ski area that could accommodate corporate jets. As a result, it had rather modern and extensive facilities for such an out of the way airport. The Eagle FSS was run by an old fellow who told us that he had operated the airport at Rifle in the 1940's and 50's. He had also flown Cubs in this part of the world for many years. One of his stories was that they routinely landed on the highway, taxied into a gas station and fueled up. Try that today and there would be a hundred cops on the scene.

I told him that we were headed to Kremmling, Colorado and were worried about the high country. He gave us some hints as to how to fly the valley around Eagle and where to gain some altitude. After we took off, I contacted him on the radio. He watched us out his window. He would tell us to fly to a particular spot alongside a hill as it usually had some lift and he was right. After about 10 minutes of this the Cub had gained a lot of altitude and we headed east up the river valley. I had planned to turn north at the village of Dostero and then follow the Colorado River. He suggested we fly up to Wolcott and turn north when we saw the red cliffs. Ann and I did and turned into a scene right out of the beginning of the movie <u>The Sound of Music</u>. There were large well-rounded hills below us that stretched out to both the right and left finally merging with the higher mountains. With the spring's new grass, the treeless hills were a lush medium green color. Between the hills, a highway wound its way to the north.

By this time, I had worked the Cub up to 9,000feet and Ann and I could see forever. In 10 minutes, we reached the Colorado River at the town of State Bridge and began to follow the river to Kremmling. Five miles southwest of Kremmling, the river and a railroad enter the narrow mile long Gore Canyon. From the air, it offered a somewhat lower path to Kremmling, so I headed the Cub toward it. A couple hundred yards from the mouth of the gorge, a very strong down draft was encountered. I quickly made a 180 degree turn to get out of it. Another 10 minutes were spent regaining the lost altitude. I made another pass with a higher entry, but the result was the same. An alternative route was needed. Off to our right the Kremmling VOR sat on top of a 9,400foot peak. To the right of it was a somewhat lower saddle in the mountains. It offered a better chance of getting to Kremmling.

I had lost radio contact with the Eagle FSS, but continued to transmit on the original frequency. Theoretically, I could have talked to him via the VOR, but Ann could not find a workable frequency. After working up to a higher altitude, the Cub flew over the saddle, past the VOR and out into a broad mountain valley. We had plenty of altitude and had been flying only an hour. Rather than land at Kremmling, Ann and I opted to keep on flying to the unattended airport at Grandby. We forged on and landed the Cub there an hour and 25 minutes after leaving Eagle. All-in-all it had been a very interesting day.

As Ann and I tied down the Cub, a local fellow drove up to inspect the Cub. He let us into the airport building where I called the Eagle FSS. I talked to the old fellow from Rifle again. He had received all our transmissions and was able to keep track of our progress, however, he could not raise us on any frequency. I thanked him for the good directions and told him how pretty the route had been.

Back at the Cub, we determined that the fellow ran the local Standard Oil gas station. He was a pilot and had a hobby of photographing every airplane that landed at Grandby. He located the key to the gas pumps and we were able to refuel the Cub. Then, he gave us a ride into town, so we could eat supper. Afterwards, Ann and I walked back to the gas station. I told the fellow that we were going to camp under the airplane, so we could get an early start in the morning. He said that he would tell the local policeman we were camping at the airport. Then, he gave us a ride back to the airport.

The rain-fly was set up as a tent again. Then, we climbed into our sleeping bags for the night. Although the idea of having the police check on us was a good one, every time they drove by us I woke up thinking it was someone else. As a result, I did not get a lot of sleep that night.

A year later, JoAnn and I lived at Winter Park for the ski season. One day I took our car the Standard station for an oil change. The owner kept looking at me rather quizzically. Finally, he asked me if I had been there before. I told him that my daughter and I had landed the Cub here the year before and he had been very helpful. At that he showed us his picture of the Cub.

Ann and I awoke at sunrise on Sunday to a bright clear day, a temperature of roughly twenty-five degrees, and a brisk wind out of the west. The cold front that passed us in Monticello had left several feet of new snow on the Rocky Mountains. Above 9,000feet, everything was pure white. The wings and top of the airplane had acquired a good coat of frost during the night. We turned the Cub's back to the morning sun so its warm rays would melt the frost from the wings. While waiting for this to happen, we packed up the camping gear, and ate the breakfast goodies purchased in town the night before.

As soon as the wings were dry, the biggest challenge of the trip began—crossing the Continental Divide. My plan was to go over the Continental Divide at the 11,800foot Rollins Pass just to the east of the Winter Park ski area. I had looked at this pass, from the top of the ski runs, for many years and had good idea of what was up there. The highest altitude the Cub could normally reach ,10,500feet, is called its service ceiling. Somehow we needed to fly much higher than that to cross the divide.

The airport at Grandby was the highest airport on our trip as it sat on top of a small mesa at 8,203 feet above sea level. I taxied to the east end of the 5,200foot runway that sloped slightly down to the west. There was a good sized drop off at the end of it. Ann got both cameras ready. I pushed the throttle to full power and the Cub lifted off after using only 2/3rds of the runway.

Halfway to the Winter Park ski area there is a 10,681foot peak called Sheep Mountain that stands all by itself in the Fraser valley. It is somewhat rounded and is covered with trees. I headed for its west side hoping to get some slope lift. We were now above 9,000feet and as I anticipated, there was some good lift off the west face of the mountain. In a few minutes we were at 10,000feet, so the Cub was headed across the valley toward the Continental Divide.

From our present altitude, Ann and I had a nice view of the ski area. With their coat of fresh snow, all the runs stood out. It was probably more fresh powder than they had during the whole ski season. As we headed across the valley, the Cub maintained its altitude even though we were now flying at its service ceiling of 10,500feet. From our drift angle, I could tell there was a 15mph wind directly out of the west blowing over the cold snow right up the face of the Continental Divide. At this point my experience of flying radio-controlled model gliders came into play. I had once flown a model glider in the slope lift off the roof of a barn for thirty minutes. I decided that the same technique would work here.

As the Cub approached the west face of the divide, I set the throttle at cruise RPM, initiated a right hand turn to the south, and flew parallel to the mountainside. The left wing tip was just above the trees, much the same as we had done along the Glenwood Springs canyon. After a mile and a half, a turn away from the mountains toward the valley was initiated. One of the tragic aspects of mountain flying is that pilots do not realize that the mountains go up a lot faster than a light plane can climb. The east face of the divide is littered with the remains of airplanes. Their pilots belatedly realized that they either had to turn back or fly into the mountain. When they turned, the wing stalled and the plane crashed. John and I had hiked up to several of these crash sites. In each case, there was not much of the airplane that was intact.

Being well aware of the climb problem, I wanted my turns to be out into the valley not toward the trees. If the Cub was going to drop out of a turn, I wanted plenty of airspace below it. After making a 180 degree turn, I headed back to the face of the mountains. I put the right wing tip just above the trees and then flew to the north for several miles. Another turn out into the valley was made and the figure eight completed. The Cub was gaining a good amount of altitude with each pass along the face of the mountain. I kept repeating the figure eight pattern while we slowly worked our way up to altitude.

Finally, the Cub was above tree line and we could move a bit closer to the side of the mountains. Ann was so busy taking pictures with the 35mm still camera and the 8mm movie camera that she barely had time to look at the scenery. We were three miles to the north of a 13,294foot peak and twenty-one miles south of the 14,255foot Longs Peak. Because of the deep snow cover everything looked like a winter wonderland. The sky was a deep blue and there were a few puffy white cumulus clouds scattered about. After forty-five minutes of doing figure eights, the Cub was a few hundred feet above the pass. Ann and I could look down at the sign, put up for the hikers, that named the pass and gave its history.

Now, the decision was to just sneak over the pass or get more altitude before going over. The major concern was the possibility of a downdraft as the wind went over the divide. One time, John and I had sat on top of the divide watching the clouds. The up-sloping west wind would lift moisture up over the divide where it would condense and form a small cloud. As it went down the east side the cloud would dissipate. As a result, I was familiar with the existence of the downdraft phenomena.

Since the Cub was gaining altitude rather easily, Ann and I decided to make two more figure eights before going across the divide. On the last northward pass the Cub was at an altitude of 12,300feet and 500 feet above the low spot of the pass. Ann leaned over my shoulder and photographed the altimeter to document our height.

The last time I had planned to be this high in an airplane was on a weather reconnaissance flight over the Yellow Sea. There was a route called "Charlie" that went from Seoul, Korea to Port Arthur, Manchuria, along the Shantung peninsula, and then south to the mouth of the Yangtze River near Shanghai, China. Normally, this route was flown about 20 feet above the water to avoid detection by the Chinese radar. As our B-26 approached the mouth of the Yangtze River, I took a radio fix on the Shanghai classical music station. Our crew often listened to this station on the long over water missions. The station played a mixture of Brahms, Beethoven, and classical Chinese music. When I calculated our position, we were only thirty-two miles from Shanghai and well inside Chinese territorial waters. In a few minutes, we reached the point where some genius in the Fifth Air Force weather section had planned for us to circle the B-26 up to 12,000 feet taking weather measurements on the way up.

By now we had broken out of the scud clouds and the skies were clear. I was sitting in the nose and had a spectacular view of *the Chinese coastline and the mouth of the Yangtze River. As planned, the pilot started a slow spiral climb. At 3,000 feet, a glint of silver on the ground, roughly eight miles away, caught my eye. Closer inspection revealed that the silver specks were rows of MIG-15 fighter jets at a Chinese military airfield. I quickly told the pilot what I had seen. He took one look at the rows of MIG's, racked the plane over into a steep dive and headed for the waves. We quickly reached the protective cover of some scud clouds. Then, headed straight for home base at 20 feet off the water. Back at K-14, we informed the weather office that if they wanted the weather above the mouth of the Yangtze river, they could go get it themselves. The spiral climb portion of the route was never scheduled again.*

In case there was a need to make a turn, the Cub was pointed at a 45-degree angle to the Continental Divide. As the Cub flew over Rollins Pass, the dreaded downdraft never materialized. The Cub sailed over the pass in grand style. Now, we could see at least a hundred miles out into the plains. Below us was the east end of the Moffett tunnel with its railroad switching yard. Ahead of us, glistening in the morning sun, was the Nederland reservoir. The east side of the divide has a much shallower slope to it than does the west side. I set up a nice slow descent while keeping plenty of clearance above the ground.

In five minutes, the Cub was well clear of the divide. Then a feeling of sadness swept over me as our greatest challenge was over. Ann and I had flown the Cub over the Rocky Mountains. I positioned the Cub, so that Ann could take some pictures of the rapidly receding snow capped Rollins Pass. Then, we resigned ourselves to just getting home. Since we had so much altitude, I pulled on the carburetor heat, set the throttle at 1,700rpm, and the Cub slid quietly downhill the sixty miles to Greeley, Colorado.

The whole flight from takeoff at Grandby, over the divide, and on down to Greeley took only one hour and 50 minutes of flying time. On the ground, both Ann and I had a sense of relief that we had not crashed into the mountains, but also were saddened that the challenge was gone. I told the lineboy where we had come from. He gave me a quizzical look that implied that we either were crazy or telling him a story.

From Greeley, the route of flight took us to Sterling, Colorado and then to North Platte, Nebraska. We were able to fly the 140 miles from Sterling to North Platte in two hours and ten minutes nonstop. This was pushing things a bit, but the weather was typical Colorado severe clear. In addition, the Cub was making a ground speed of 67MPH and both Julesburg and Holyoke were only a few miles from the midpoint of the flight. By now, Lee Bird Field at North Platte was old familiar ground.

Because we were going to go to Worthington, Minnesota to visit JoAnn's sister Clarice and then onto Mankato to see my parents, I laid out the most direct route. It took us to O'Neill, Nebraska, Yankton, South Dakota, and then Worthington.

It was 125 miles from North Platte to O'Neill. For this leg, Ann would be the pilot in the front seat and I would attend to the navigation duties. In the first thirty-five miles there was not a single landmark useful for navigation purposes. I had estimated a wind correction angle, so all that Ann could do was fly the corresponding compass heading. In thirty-two minutes, the small town of Arnold appeared. It was a welcome sight for two reasons: first, we were not lost; and second, our compass heading was very good.

The first 20 miles from Arnold to O'Neill possessed some of the most desolate country that Ann and I had flown over. It was mile after mile of gently rolling prairie that flowed to a distant horizon. The only sign of human habitation was an occasional windmill pumping water for the cattle. Even the cows were widely scattered over the countryside as if even they did not want to have much to do with each other. In this part of the world, even a ranch house gets a symbol on the map, but we never did see one.

From a navigation point of view, the Arizona desert was a much friendlier place as there were mountains, canyons, and all sorts of things to tell you where you were in the larger sense. Here, the countryside appeared the same regardless of the direction you looked. All we could do was: hold our compass heading, compute the ETA to the next check point, and hope the wind did not shift. We had to be sure not to miss what few landmarks there were. Fortunately, Ann was in the front seat as she can hold a compass heading much better than I.

Finally, we crossed the highway and railroad track that connects Theford and Broken Bow. Then, we crossed the Middle Loup River, the North Loup River, and finally the Calamus River. From there to O'Neil it was more prairie with only a few roads for check points. After two hours of flying through truly what was "nowhere" Ann had O'Neill in sight. I'm sure that if we had been flying at a thousand feet above the ground, we could have simplified the navigation, however, we were at our usual 100-200 feet. We landed at O'Neill and refueled the Cub.

From O'Neill to Yankton, South Dakota was only 71 miles. Once in the air, The Cub quickly reached the valley of the Missouri River and the Lewis and Clark Lake. Then, we flew over the north end of the two mile long earthen dam that creates the lake. When putting gas in the Cub at Yankton, the lineboy told us that today was the first time in a month they had aviation gasoline. This was the period of the 1973 Arab oil embargo and gasoline was in short supply. Strangely, this was the first airport where there was any question of fuel availability. Happy to have landed on the right day, I paid the bill and we took off for Worthington, Minnesota--a hundred miles away.

Now, the Cub was flying over some of the flattest farmland in the Minnesota. There is mile after mile of nothing but corn and soybean fields. After a day of flying over the grey-green range land, the lush green fields now beneath us were a welcome sight. It also was a lot warmer. The rectangular pattern of the roads makes it easy to keep a compass heading and there are enough small towns to keep track of where we are. It took an hour and 40 minutes to fly the hundred miles. Ann and I stayed over night with Clarice and Frank and told them the highlights of our trip. Monday morning , June 4 1973, we took off for Mankato to visit my parents. As usual, we arrived unannounced to spend the night with them.

The next day, Ann and I got a rather late start getting out of Mankato and on our way to La Crosse, Wisconsin. As the Cub approached La Crosse, a black line of thunderstorms developed to the south. Even though it was getting dark, they seemed to be off in the distance. We landed and refueled the Cub. The lineboy was a gentleman, about 65 years old, who asked us where we were heading. I told him:

"Madison."

He replied:"You had better check the weather at the FSS."

Suffering from get-home-itis, I told him that I thought that we could keep ahead of the storms. Ann and I took off and headed for St. Joseph. As the Cub went past the convent, the weather collapsed. It started to rain, the wind began to blow, and it became very dark. I realized that we were not going to be able to outrun the storm. The old guy had been right.

Fortunately, I was in the front seat where it is easier to cope with tough conditions. I turned back toward La Crosse and in a few minutes the Cub was on final approach to runway 22 that pointed right into the very strong wind. I made a wheel landing and once on the ground the wind tried to lift us off again. At about 50% power and holding lots of forward stick pressure, I could keep the wheels on the ground and proceed up the runway.

When we reached the taxiway, it was necessary to taxi the Cub all the way to the ramp in a strong 40 degree crosswind. The rain was coming down in wind driven sheets, it was very dark, and an occasional flash of lightning would light up the taxiway ahead of us. I was holding the stick full to the upwind side and using the rudder to maintain the heading. By now, I was carrying nearly 75% power and when a wind gust hit us the Cub would go airborne. I would push the stick forward, give a jab of power, kick the rudder, and get the Cub in a one wheel landing configuration. With one wing low, lots of power and down elevator, the Cub would again make contact with the ground. Then the gust would be past us, I had to partially undo the control inputs, pull back on the throttle, and the Cub would be rolling on just the upwind wheel. Another set of corrections and both wheels would be firmly in contact with the ground. As soon as things were straight, the next gust would hit us and I had to start our tiptoe routine all over again.

In the midst of all this, I told Ann that we were going to taxi up to the FSS side of the airline terminal and she would have to tie the Cub to the chain-link fence. To do this, Ann needed to retrieve the tie down ropes out of the baggage compartment. This necessitated her undoing her seat belt, turning around, putting most of the baggage and camping equipment on the floor while all the time the Cub was doing its pirouette. With her usual efficiency, she had the ropes ready in thirty seconds.

I taxied up to the fence. Ann opened the door, jumped out into the storm, and quickly tied the right(upwind) wing strut to the fence. In the meantime, I was playing the throttle, ailerons, rudder and elevator to keep my position relative to the fence. With one wing tied down, Ann ran around the back of the plane and then tied the left wing to the fence. There was nothing to tie the tail to, but we were pretty much into the wind. I killed the engine, waved Ann off, climbed out of the Cub, and we both ran into the FSS.

One of the FSS men said that they had been watching us and the wind velocity meter as the Cub taxied up to the building. He said that the gusts measured 52 mph. He was amazed that we were able to taxi crosswind for about two city blocks in such a wind. A week earlier a Citabria pilot had tried to do the same thing. The wind flipped that airplane on its back. In my opinion, the Cub is one fine crosswind landing airplane. Today proved that, with a bit of luck, it could handle a lot.

In twenty minutes, some men, from one of the corporate hangars, came down to help get the Cub out of the storm. As they hung onto the wings and tail, Ann and I untied the Cub. The group of us walked it 200 yards along the ramp, then through the open door, and into a large hangar. I called JoAnn and told her to drive to La Crosse to get us. After two and one half hours of driving through rainstorms she showed up and we drove to a local motel. JoAnn did not have any overnight things and I was low on funds, so three of us drove back to Madison. I must have been a bit hyper from all the excitement as we should have stayed in La Crosse that night.

The next morning(Tuesday), I told Ann that her mother and I would drive up to La Crosse, so I could retrieve the Cub. Ann would hear none of that. She wanted to complete the whole trip in the Cub. When we arrived, the Cub was sitting out on the ramp and not tied down. I was unhappy about this as a corporate jet or a twin blasting its throttles could have easily flipped the Cub. As often is the case after a line of thunderstorms goes through, the following day was clear and calm. Ann climbed in the back seat, I flipped the prop and we were ready to taxi out.

After takeoff, the Cub climbed along the course line and passed over the convent, at St. Joseph, where we had lost the race with the storm. Today, everything was serene and pastoral under blue skies and a bright sun. In a few minutes, the Westby ski jump came into sight. After that we could see Blue Mounds forty miles off to the right and the Baraboo bluffs on the horizon to the left. Since there were no navigation problems left, The Cub just coasted along at one hundred feet above the tree tops and Ann and I enjoyed the scenery. When the Cub touched down at Waunakee, it had been 28 hours and 50 minutes of flying time since we had taken off from Prescott, Arizona. Our average ground speed, for the 1,604 miles was 56 mph, which was a bit better than the 50 mph on the way to Prescott. Overall, the 3,247 mile trip took 61 hours and 50 minutes of flying time.

Postscript

Six of the pictures that Ann took on the trip are now on my office wall and also on the wall of her apartment. When I get tired of working, I put my feet up and mentally relive the flight mile by mile. The movies also are wonderful mementos. It thrills me each time I see the Grand Canyon, Monument Valley, or watch us slowly fly figure eights up the west face of the continental divide in the Rocky Mountains. You could take this same trip in many other airplanes, but it would never match the experience of having done it in a Cub.

Flight Log

DATE 19 73	FLIGHT FROM	FLIGHT TO	AIRCRAFT MAKE AND MODEL	AIRCRAFT CERTIFICATE MARK	CLASSIFICATION			DURATION OF FLIGHT	
					SEL			405	40
25 May	Waunakee	Dubuque Ia	Piper J-3	88354	1:45			1	45
25 May	Dubuque Ia	Marshalltown Ia			2:35			2	35
25 May	Marshalltown	Audubon Ia			2:20			2	20
25 May	Audubon Ia	Fremont Neb			1:35			1	35
25 May	Fremont Neb	Gothenburg Neb			4:05			4	05
26 May	Gothenburg	Julesburg Colo			3.00			3	00
28 May	Julesburg Colo	Sterling Colo			1:00			1	00
28 May	Sterling	Jeffco			1:40			1	40
29 May	Jeffco	Pueblo Colo			1:55			1	55
29 May	Pueblo	Raton N.M.			1:50			1	50
29 May	Raton	Las Vegas N.M.			1:50			1	50
THE RECORD ON THIS PAGE IS CERTIFIED TRUE AND CORRECT:					23.35			23	35
PILOT ______ ATTESTED BY ______					CARRY TOTALS FORWARD TO TOP OF NEXT PAGE				

DATE 19 73	FLIGHT FROM	FLIGHT TO	AIRCRAFT MAKE AND MODEL	AIRCRAFT CERTIFICATE MARK	CLASSIFICATION			DURATION OF FLIGHT	
					SEL			429	15
29 May	Las Vegas	Albuquerque	Piper J-3	88354	2:00			2	00
29 May	Albuquerque	Grants N.M			1:30			1	30
29 May	Grants	Gallup			1:15			1	15
30 May	Gallup N.M	Winslow Ariz			2:25			2	25
30 May	Winslow	Prescott Arizona			2:15			2	15
31 May	Prescott	Grand Canyon			1:30			1	30
31 May	Grand Canyon	Page Ariz			1:55			1	55
1 June	Page Ariz	Oljato Utah			2:10			2	10
1 June	Oljato	Blanding Utah			1:05			1	05
2 June	Blanding	Mohab Utah			1:45			1	45
2 June	Mohab Utah	Grand Junction Colo			1:25			1	25
THE RECORD ON THIS PAGE IS CERTIFIED TRUE AND CORRECT:					19.05				
PILOT ______ ATTESTED BY ______					CARRY TOTALS FORWARD TO TOP OF NEXT PAGE				

DATE 19 73	FLIGHT FROM	FLIGHT TO	AIRCRAFT MAKE AND MODEL	AIRCRAFT CERTIFICATE MARK	CLASSIFICATION			DURATION OF FLIGHT	
					SEL			448	20
2 June	Grand Junction	Eagle Colo			1:40			1	40
2 June	Eagle Colo	Granby Colo			1:25			1	25
3 June	Granby	Greeley Colo			1:50			1	50
3 June	Greeley	Sterling			1:10			1	10
3 June	Sterling Colo	North Platte Neb			2:10			2	10
3 June	North Platte	O'Neil Neb			2:00			2	00
3 June	O'Neil Neb	Yankton S.D			1:10			1	10
3 June	Yankton S.D	Worthington Minn			1:40			1	40
4 June	Worthington	Mankato			1:40			1	40
4 June	Mankato	Lacrosse Wis			2:45			2	45
5 June	LaCrosse Wis	Waunakee			1:30			1	30
THE RECORD ON THIS PAGE IS CERTIFIED TRUE AND CORRECT:					19.00				
					CARRY TOTALS FORWARD TO TOP OF NEXT PAGE				

On The Road in the Navaho Indian Reservation

Monument Valley

Rollins Pass

12,300 Feet

Chapter 6
"Cubbing" With JoAnn

A complaint of many pilots is that their wives do not participate in their flying activities. Fortunately, JoAnn is not one of these wives as we do considerable flying together. Numerous trips have been made to southern Minnesota in order to visit either her family or my parents. Often the weather on these trips was not the best, but JoAnn accepts it as part of flying. Much to my delight, she is absolutely unaffected by rough air and turbulence. In fact, she handles it better than I do.

Fly-In Breakfasts

JoAnn and I often participate in an institution that is almost as old as aviation itself--the Fly-in breakfast. The standard menu is pancakes, sausage, scrambled or fried eggs, and milk or coffee. Fly-in pancakes are some of the best in the world and are a gastronomic delight rivaling those of Paris. On the day of a Fly-in breakfast, pilots within an easy flying distance arrive early to be there before the food line gets too long. Later in the morning, the local townspeople arrive after church and the time spent standing in the line can get quite long. Those who flew in spend the rest of the morning walking around looking at airplanes, talking to people they know, and occasionally meeting someone new. This "hangar flying" is a very valuable part of aviation. One learns about the kinds of problems people are having with their airplanes, who is restoring a particular airplane, and what evils have befallen aviation recently.

In the Madison area there are about ten Fly-in breakfasts that are held regularly each summer. The Rio Fly-in is our favorite as the food is excellent and pilots from many parts of Wisconsin are regular attendees. The Rio airport is an east-west grass strip with pine trees on the west end and a tall Cottonwood tree at the east end. On a hot summer day, taking off to the west in the Cub can be a questionable affair as the Cub's climb is notoriously slow under any conditions. Every once-in-a-while, I chicken-out before the Cub gets to the trees and turn left to avoid hitting their tops with the wheels. Taking off to the east requires an early left turn to avoid the Cottonwood tree, but I then face a major power line along a raised railroad track. On a hot day, the power line could be an obstacle, but fortunately it dips just about where the Cub needs an extra ten feet of altitude. Despite these obstacles, there never has been a takeoff or landing incident at a Rio Fly-in breakfast.

The Sextonville Airport Restaurant

Several times each summer JoAnn and I fly to the Sextonville airport for Sunday lunch. The airport's official name is the Richland Airport, however, everybody calls it Sextonville after an adjacent hamlet. To get to this airport, we fly the Cub due west from Waunakee. Some 15 miles after takeoff, we can see where the Wisconsin River enters a broad valley that goes all the way to the Mississippi River. A few miles before the village of Spring Green, a highway crosses the Wisconsin River on a picturesque multi-span steel truss bridge. To the northwest of the village is the un-glaciated area that stretches to the Mississippi river. As mentioned in earlier chapters, this country is a maze of ridges and valleys with numerous creeks wending through the valleys on their way to the Wisconsin River. Eleven miles due west of Spring Green, one of these valleys cuts through the Wisconsin River's north bluffs. A few miles up this narrow valley the airport is nestled against the east slope of the valley. It has a 2,100 foot north-south grass runway, and a very short east-west grass runway. The latter has several tall cottonwood trees on its east end that reduce the effective length of the runway. Locating other airplanes can be a problem as on final approach the planes fly low in the valley and are difficult to see against the hills.

The Sextonville airport is famous for its on-the-field restaurant in a small metal building. It can seat about 50 people and is run by two elderly women, Hilda and Betty. The FBO here, who gave me the valuable advice on my first solo cross-country in the Cub, is Betty's brother Al. People fly in from all over the Midwest to have Sunday lunch here. The fellows from the Stoughton, Wisconsin airport put Norwegian flag decals on their airplanes and called themselves the Norwegian Air Force. On given Sunday, they would stagger their departures from Stoughton, so that they all would arrive at Sextonville at the same time. The airport is also only about 50 miles from Camp Mc Coy where National Guard training takes place in the summer. On many Sundays, there will be several military helicopters arriving at once which is an impressive sight to see and hear.

When describing the restaurant, I tell people that it is just like going to someone's grandmother's farm house for Sunday dinner. The menu is always the same: Waldorf, Jello , and bean salads; stringy beef pot roast, chicken, roast pork, mashed potatoes; and wonderful home made raisin bread. Dessert is always a variety of homemade pies such as banana cream, coconut cream, and in season either strawberry or raspberry pie. The food is served cafeteria style. I always vow to not take too much food, but invariably pile my plate too high. In the summer, there are picnic tables outside where JoAnn and I can watch the airplane traffic while we eat. The restaurant also draws a clientele of elderly people from the nearby towns and farms. Some of the "regulars" are there so often that it seems like we know them even though we do not.

One Sunday, JoAnn and I were going to take off to the west from Sextonville when I saw a twin engine Cessna 337 fly over going to the south. I thought it was going to land to the north, waited a few minutes, but it did not reappear. Assuming it was just passing by, I pushed the throttle forward to start my take off roll. Just as I did, the 337 went by about 25 feet above the Cub and landed not too far down the runway. If I had started rolling a few seconds earlier, it would have landed right on top of us. Later, I talked to the pilot, he said that he saw us sitting on the east-west runway and assumed that it was the active runway even though it was very short. It turned out that the Cessna had a STOL conversion. The short runway and tall trees gave him a chance to show off its capabilities. Sitting on the runway, it never entered my mind that a 337 could land on the east-west runway.

As JoAnn and I left the Sextonville airport on another Sunday, I decided to fly to the south side of the Wisconsin River valley and follow the bluffs back towards Madison. Upon reaching the south side, I turned to the east and was flying the Cub about 75 feet below the top and rather close to the bluffs. There was a road about a third of the way up the bluffs that followed the river. I was busily watching a pick-up truck going along the road below when JoAnn yelled: "WIRES." I switched my gaze from the truck to out front just in time to see a set of power lines sweep by about six feet below the wheels. It happened so fast I didn't have time to even think about pulling up. What saved us from hitting the wires was that the pick-up had gone up a slight rise in the road and to keep my altitude above the truck I had subconsciously pulled back on the stick. After getting over the shock, I made a 360 degree turn to recheck the scene. There was a power pylon hidden in the trees at the top of the bluff and another one on the valley floor across the river also in a woods. The power lines went down at a 45 degree angle across the river and were nearly invisible against the trees. That incident made me ever vigilant for power lines. It also cured me of watching objects on the ground for more than a few seconds at a time while flying low.

Visiting Ann

Although JoAnn and I do considerable local "airport hopping", we also use the Cub for cross-country flights. A prime destination for a Cub trip was Midland, Michigan where our daughter, Ann, was a research chemist at the Dow Chemical Company. There are two possible routes to Midland. One goes south around the Chicago area and then north to Midland. The second goes around the north end of Lake Michigan and then south to Midland. The latter route was chosen as it was through a part of the country that we previously had not flown over. The distance from Madison to Midland is 497 miles. To go from upper to lower Michigan the Cub would have to fly across the five miles of open water at the Straits of Mackinac. A task that would elicit a high "pucker factor".

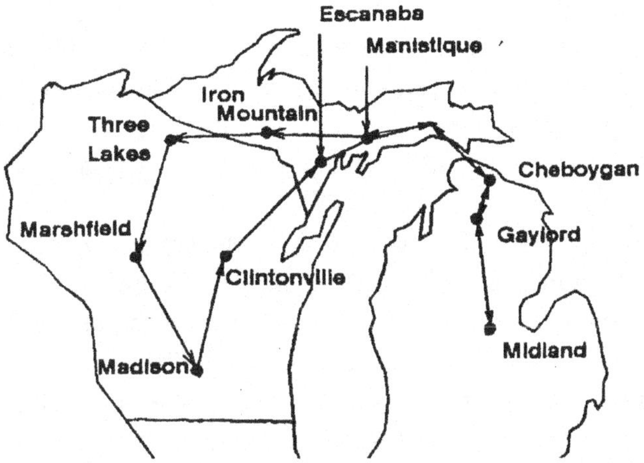

Route of Flight: Madison to Midland Michigan

JoAnn and I left Waunakee, early on Friday, June 6,1975, with the Cub pointed a few degrees to the right of north. The first gas stop was at Clintonville, Wisconsin. The next airport on the route was at Escanaba in Upper Michigan--110 miles away on a heading of 45 degrees. On final approach to the Escanaba airport, I could see a North Central Airlines Convair 540 on the taxiway at the midpoint of the runway waiting to get onto the runway. Knowing that airline pilots like to keep on schedule, I thought we would be helpful by getting out of the way as soon as we landed. Once on the ground, I taxied the Cub off the runway into knee-deep grass and waited for the Convair to go by. After it went past, the Cub was turned around to get back onto the runway. Suddenly the right wing dropped, the propeller started throwing chunks of dirt and grass. As soon as the dirt flew, I turned off the engine. After the right wing tip touched the ground, JoAnn and I climbed out of the Cub to survey the situation.

The Cub's right wheel had dropped into a hole about two-feet in diameter and four-feet deep. The hole had been dug next to a runway light in order to repair the light's wiring. After about 10 minutes, some airport workers came out and lifted the Cub out of the hole. Despite all the dirt that was slung about, the propeller was not broken. There was a good dent in the brass leading edge on one tip and a small one in the other tip. After inspecting the propeller carefully, I decided that it was safe to continue flying.

All during the flight to Manistique, in upper Michigan, I kept hearing strange sounds coming from the prop and was worried about it tearing apart. My concern was heightened by our flying over pine forests and cedar swamps with no place to land the Cub. At Manistique, the airport's mechanic look at the propeller. He thought that there was no real problem with it. The strange noises were a combination of a bit of distortion of the prop tip and my hypersensitive hearing. While we were there, he showed us a very large collection of A-65 engines and parts that he was in the process of selling. I wished I could have afforded to buy them. Later, I heard that someone from California bought the whole lot for $2,000.

Initially, the 130 mile flight from Manistique to Gaylord, in lower Michigan, was along the north shore of Lake Michigan. JoAnn and I were flying over swamps, pine forests, and occasional fields full of stumps. The shoreline of Lake Michigan was a mile or so off to our right. Due to the medium height grey overcast, it was a rather gloomy day. Despite this, the visibility was excellent. As the Cub approached the Straits of Mackinac, I flew off to the side of the five mile long bridge and high above it. If the Cub went down, I wanted people in cars on the bridge to see it happen and get help for us. The thought of forced landing in the lake did not appeal to me. If the Cub is landed on the water, it will flip on its back and its occupants will be hanging upside down from their seat belts under water. Getting out of a submerged and upside down Cub was not a sure thing.

As the Cub reached the eastern shoreline of Lake Michigan, the height of the gas gauge wire indicated that it would be prudent to refuel. We deviated from the planned line of flight and landed at the Cheboygan, Michigan airport.

Between Cheboygan and Gaylord there is a range of high hills with a major highway going up and over them to the south. As we approached the hills, I put the Cub in its climb mode. Just about the time the Cub was high enough to get over the hills, it was in the clouds. A northwest wind was pushing moist air off the lake, up the range of hills, and then clouds formed. In the winter, this same lake effect dumps large amounts of snow on these hills. Near the top of the hills was a large resort with a grass airstrip. I could get to the airstrip, but the minute we went past it the Cub was in the clouds again. After ten minutes of poking around the hills, it became obvious that the hills were impenetrable, so we flew back to Cheboygan and stayed there overnight.

In the morning, the skies were clear, so JoAnn and I took off and headed the Cub south. At Gaylord, a fellow in the pilot's lounge, offered to show us his Fairchild 24 that had won a grand prize at the EAA convention in Oshkosh the previous summer. Unfortunately, I was in a hurry to get to Midland and declined his offer. By mid-afternoon, we landed at Midland, refueled the Cub, tied it down on the ramp, and waited for Ann to pick us up.

Midland was a small city with very large chemical plants devoted to processing caustic soda. The fly ash from these plants is quite corrosive. It will pit the paint on automobiles and eat away exposed metal. The houses had marble window sills to prevent such corrosion. Ann gave us a tour of the city that included a beautiful Civic Center where an art fair was being held. Because of many such facilities that were not normally found in a city of Midland's size, it was an excellent city in which to raise a family. Ann said that it was not a very exciting place for a young adventurer to live.

After two days with Ann, it was time to return home. Early Monday morning, the Cub was headed north. Before the Straits of Mackinac, we stopped at Pellston, Michigan for fuel. This stop also shortened the leg to Manistique to 92 miles. After going across the straits, The Cub was flown, on a westerly heading, about a mile inland of the Lake Michigan shoreline. Roughly 20 miles east of Manistique, a dense layer of low fog was encountered that covered the ground for several miles inland. The Manistique airport was on low lying ground only a mile from the lake shore. To compound matters, the Cub was past the point-of-no-return to Pellston. The next airport on the course line was well beyond our range. I also knew that in this land of pine forest, cedar swamps and fields of stumps there was no place in the vicinity for an off-airport landing. The Cub, JoAnn and I were in real trouble.

With considerable trepidation, I set up a long straight in landing approach to where I thought the Manistique airport was located. As the Cub came close to the airport, I was surprised to see that the runway was clear of fog. But at its sides, the fog formed a 75-foot vertical wall. Overjoyed, I kept on coming down on final approach. As the Cub landed, I told JoAnn that I felt just like Moses parting the Red Sea.

Once off the runway, the Cub was taxied through dense fog to the ramp. The Cub was tied down and then JoAnn and I walked to the airport office. The same fellow who had inspected the Cub's propeller was there again. He was astonished to see us. He asked us where we came from. I said "Pellston in a Cub." He could not believe that we had landed a Cub here. He said a Mooney 21 had been circling for an hour trying to land and had given up. I led him on for a bit and finally told him that the sun's heat had dissipated the fog over the black asphalt runway. I was not sure he believed me as the Cub, parked nearby on the ramp, could not be seen. In another 20 minutes, the fog was gone. JoAnn and I went on our way, leaving the fellow quite perplexed.

From Manistique we flew to Iron Mountain, Michigan where a recent hangar fire had destroyed a number of airplanes. The 47 miles from Iron Mountain to Three Lakes, Wisconsin was across some of the most desolate country I have ever seen. It was mile after mile of cedar swamps, pot hole lakes, scrub pine trees and an occasional dirt road that seemed to go nowhere.

Once past this area, JoAnn and I stopped at the Three Lakes airport to check out a campground adjacent to the airport. We talked to the woman who ran the place about airplane camping there. She had no problem with our tenting under the wing and using her facilities as long as we paid a camping fee. We kept that in mind for the future.

The next leg was to Marshfield, Wisconsin. En route, the countryside beneath us was much more hospitable than that of the previous leg. After refueling the Cub there, we took off and took up a southeasterly heading to avoid a restricted area southwest of the city of Wisconsin Rapids. The National Guard fires artillery pieces in this area and the shells go high in the sky---no place for a Cub to be. We flew over Wisconsin Rapids and then took a more southerly heading. From there to the Baraboo Bluffs, the countryside is not very navigation friendly. There are many areas of swamps, a few farms with small fields of corn or alfalfa, and only a few roads. At the half-way point there is a tall isolated ridge that has a ski area on its west side. Other than this, there is not much to see on this 105 mile leg to Waunakee.

Airplane Camping With JoAnn

Washington Island

JoAnn and I also use the Cub to go camping in northern Wisconsin. Getting all the camping equipment into the narrow confines of the Cub was a feat. A backpack frame was strapped to the cockpit's overhead tubes. The tent and sleeping bags were stuffed on top of the baggage compartment lid. A small backpack containing our clothes was placed in front of all this and tied in place. The cooking equipment and food were placed in bags on the floor in front of the back seat. Once in the back seat, JoAnn was in very tight quarters.

On August 16, 1973, we flew the Cub to Washington Island, a large island in Lake Michigan, just off the tip of the "thumb" of Wisconsin. This entails flying over five miles of open water, which does not thrill me at all. About halfway across, there is an island with a light house that provides some small amount of comfort.

After landing on the grass runway at the Washington Island airport, the Cub was tied down. The camping gear was placed on my backpack frame and JoAnn carried the remainder. We hiked the mile to the School House Beach campground. It is in a cedar forest on a half-moon shaped beach that is covered with smooth rocks about the size of goose eggs. Since the water is very cold, swimming is limited to the brave on a hot day. JoAnn and I set up the two-man tent, cooked our dinner over our small propane stove, and camped overnight.

At dawn, the sun's rays filtering through the trees and the wood smoke from the campfires gave the place a surrealistic look. All-in-all, it was a very nice place to camp.

Walking back to the airport, a car stopped and offered us a ride to the airport. The driver was a well-known collector of WW II fighter planes. On the way to the airport, we had an interesting chat about his owning and flying "warbirds."

After re-packing the camping gear in the Cub, JoAnn and I took off and headed back toward Madison. Due east of the city of Green Bay, the Cub suddenly started to shudder violently. My first reaction was wonder what had gone wrong. Then, I immediately pulled the power back to idle. This reduced the shaking to an acceptable level. I looked around, saw a recently mowed hayfield, and landed the Cub in it. After shutting down the engine, we could see that the outer six inches of a propeller blade's brass leading edge had departed taking a considerable chunk of wood with it. My first thought was that I was glad that the propeller had failed now rather than when Ann and I were flying over the Rocky Mountains.

This Hartzel propeller had been purchased at the EAA Fly-in at Rockford, Illinois. It had been sent to Whirlwind Propellers to be inspected and refinished. When I got it back, I asked them how long they thought it would last and the reply was "about a year." They were correct as it was almost a year to the day.

The farmer, on whose field we had landed, drove us to Burr Fontaine's mother's apartment in Green Bay. She graciously invited JoAnn and I to stay overnight with her. The next day Chris arrived with my old high pitch Lewis propeller. The three of us drove back to the farm and I installed the propeller. To keep Chris company, JoAnn rode home with him in the car, while I flew the Cub home alone. Once home, I bought a brand new Sensenich 72-42 wooden propeller.

A few years later, JoAnn and I flew back to Washington Island and camped there over the fourth of July weekend. The night was ruined by some rowdies that drove through the campground making trouble. Because of a series of problems like this one, the campground was closed for good at the end of the summer.

Madeline Island

In August of 1976, JoAnn and I flew the Cub to Madeline Island, one of the Apostle Islands in Lake Superior that are off the northern shoreline of Wisconsin. The route of flight was Waunakee-Stevens Point-Rhinelander-Boulder Junction-Ashland-Madeline Island. The Stevens Point airport is called Central Wisconsin and has airline service. The east-west main runway was concrete and very long. Having had some practice, I was getting better at landing on such runways.

From Stevens Point to Rhinelander the line of flight went five miles east of Wausau and its Rib Mountain ski area.

Twenty miles further on was John Hatz's private airstrip where I once had attended a Fly-in. He is the designer of the Hatz biplane that is popular with homebuilders in the EAA. He also was well known for sleeping under the fuselage of his airplane at a Fly-in. When evening came, he would wrap a blanket around himself, roll under the plane, and go to sleep. Once, two lawyers landed a twin engine airplane at his strip to inform him that an airplane he had worked on had crashed. They asked him how much insurance he carried on his work. His reply was "none." The two lawyers looked at the airport's well worn buildings, climbed back in their plane, and departed. John never heard from them again.

The busy Rhinelander airport was a cut above the typical Wisconsin airport. It had nice looking buildings and a number of airplanes parked on the ramp. From Rhinelander to Boulder Junction, the Cub was flying over the heart of Wisconsin's resort country. There seemed to be hundreds of lakes, many ringed by summer cottages. The Boulder Junction airport is open only during the summer, but it has fuel. The runway was a reasonably long, sandy, grass strip surrounded by pine trees. After refueling, we took off on runway 23. As soon as the Cub was airborne, it was over one of the ever present lakes and we were on our way.

Initially, all went well on the short 67 mile leg from Boulder Junction to Ashland. About halfway to Ashland, JoAnn came on the intercom and said:

"I have been looking at the same farm house for a long time. We not are going very fast."

I knew there were no storms in the area, so I was not concerned about the weather. None-the-less, I timed a pair of checkpoints. The computed ground speed was only 12 mph. JoAnn and I had a quick discussion as to whether it was better to forge ahead or go back to Boulder Junction. We decided it was better to go ahead, rather than back, as the high winds would be moving east. The Cub slowly inched its way to Ashland. I landed the Cub in a 35 mph crosswind and then taxied up to the gas pit.

The lineboy came out and said:

"You should have been here a half hour ago the wind was 55 mph"

A narrow dry front had gone through very rapidly. It's only impact was high wind at it's leading edge. By the time we had completed the usual tasks, the wind was almost calm.

Soon after takeoff we overflew the village of Bayfield, a pretty waterfront village perched on a hillside. It has a world class inn where meals are served by waiters in tuxedos—too high class for a Cub crew. From our lofty perch over Bayfield, JoAnn and I could see the Apostle Islands National Lakeshore. It is composed of a number of low mostly wooded islands that are quite close to each other. The dark blue water was calm and there were low post-frontal white cumulus clouds hovering above the green islands. We flew across a few miles of open water, and then landed the Cub at the unattended Madeline Island airport.

The county campground was three miles from the airport. There was no ground transportation, so JoAnn and I strapped on our backpacks and started walking. In a few minutes, a lady, who lived on the island and worked in a condominium complex, stopped her car and gave us a ride to the county campground. It was located, on a hillside, in a pine forest. Below the hill was a half-moon shaped white sand beach that was at least 2 miles long. It was the prettiest beach I had seen since Tuman Bay in Guam. The only things lacking were the palm trees. Unfortunately, Madeline Island is in Lake Superior. Even in August, the water is too cold to go swimming. None-the-less, it was enjoyable just to sit on the sand and admire the view.

After dinner, our bag of food was hung from a tree limb six feet above the ground. During the night, I heard some noise and shined my flashlight on a group of racoons trying to get at the food. From my vantage point, the racoons, with their black eye masks, looked like a gang of bandits. Not being able to grasp the food, they eventually wandered away looking for easier pickings.

The young couple in the adjacent campsite also were from Madison. The next morning, I traded them airplane rides around the island for a ride to the airport. It was a perfect day for sightseeing. The sky was filled with rows of small cumulus clouds, the dark blue water was calm, and the visibility was unlimited. I took each of them for a short Cub ride that included a view of the magnificent beach where a lone sailboat was anchored about a hundred yards off shore.

The trip home was uneventful as the beautiful weather held. The route of flight was from Ashland to Eau Claire to Marshfield to Waunakee.

These two trips were not the end of our camping involving the Cub. Every weekend after Labor Day, the Midwest Antique Airplane Club holds its Grassroots Fly-in for its members. Most of those attending the Fly-in camp in tents under the wings of their airplanes. Initially, JoAnn and I would fly the Cub there and set up our two-man mountain tent alongside the Cub. After a few years, this tent proved to be too confining for a weekend of camping. We purchased a large tent with a ten foot by ten foot floor and tall enough for us to stand up inside it. JoAnn outfitted the tent with: two cots, a wool blanket on the floor, baskets for our clothing, and a battery powered lamp hanging from the ceiling. All this camping equipment was too bulky and heavy for the Cub, so JoAnn brings it to Brodhead in the car while I fly the Cub there. The tent would be set up next to the Cub. The tent was so large compared to the other tents that it became known as the Taj Mahal. This tent served us well for a number of years, but met its demise in on a Friday. About 7:00 P.M., a severe thunderstorm could be seen approaching the airport. I hurried to the tent and went inside. I wanted to be able to watch the Cub in case it got into trouble. In the midst of the storm there were very strong winds, heavy rain, thunder, and lightning. Suddenly, the upwind side of the tent ripped open and the wind inflated the tent like a balloon. In a split second, the tent exploded and was ripped to shreds leaving me sitting on a cot in the downpour. John's Tri-Pacer was just downwind of the Taj Mahal. I was worried that the tent poles and other debris would damage his airplane. I grabbed one of the aluminum poles still attached to the remains of the tent and hung on to it until the storm was past. Needless-to-say, the beds, clothes, and myself were soaking wet. Later, I thought less of my sitting in an electrical storm holding onto a ten foot length of aluminum tubing. The Taj Mahal was replaced, so we are ready to camp again.

Chapter 7
The Green River Wedding

Following in his sister's footsteps, my son, Chris, started taking flying lessons in the spring of 1975. His flight instructor was the same one who had taught Ann. Chris soloed in a Cessna 150 in about 6 hours and in keeping with the family tradition it was on a day with a rather nasty gusty wind. He made three pretty landings, climbed out, and gave us a big smile. As was the case with Ann, he quickly built up time in the Cub and was ready to start cross-country flights. I decided to teach him how to navigate at low level on a trip to Decorah, Iowa for fuel and then on to Rochester, Minnesota. On the Madison to Decorah segment, I showed him how to identify the small creeks and hill sides on the map and to follow the course line closely. After a few minutes of instruction, he took over both the navigating and piloting. Once across the Mississippi River it is only about 30 miles to Decorah where Chris split the runway right at his ETA. A nice piece of navigating. A week later, he went with the instructor on a dual cross-country to LaCrosse. When Chris got home, he told about having a difficult time navigating as from 2,500 feet he could not find the ridges and creeks we had used.

On his solo cross-country in the Cub, he flew to Sturgeon Bay to spend the night with the Fontaine's at their cottage on Green Bay. He called us before he left to come back home and we were waiting at the airport when he landed. As he taxied up, we could see that he was eating a box lunch that Mrs Fontaine had prepared. I asked where he had flown and he described a number of places that I knew were not on his course line. When asked about this, he said they looked interesting so he deviated off course to take a look at them. Clearly he was not a chip off his father's A to B, with no deviations, navigator block. Early in the summer, he got sidetracked and never did finish his flight program.

In August, 1975, my brother, John, was scheduled to go on a white water rafting trip down the Green River in Utah with a commercial rafting company. Being a doctor, he had been going on their trips to handle medical emergencies. Chris and I were invited to go on the trip and we quickly accepted the invitation. Chris was to spend a month backpacking in the Rocky Mountains and would meet me in Colorado. Thus, I would have to fly to Denver alone, but he and I would fly together on the return flight.

The Cub and I departed Waunakee at 8:00 A.M. on the 10'th of August and headed towards La Crosse on the "southern" route to Denver. Even though I was still flying just above the trees, the same hills, valleys and farms kept showing up when they should. Some 23 miles before La Crosse the course line goes just south of the village of Cashton. This area contains a large number of Amish farms. The farmers work their fields using horses and drive horse pulled buggies along the highways. When I fly close to one of their farmhouses, the children hear the Cub and will run out to wave at it. I give them a wing waggle and wave my arm. From the expressions on their faces, they seem to enjoy getting a response from the little yellow airplane in the sky. Often they will run across a field to keep the Cub in view as long as possible.

Earlier in the summer, I had installed a right angle drive on the tachometer, so that its cable would not rub on the gas tank. After installing the drive, the tachometer cable would break after a few weeks of use. I had replaced it three times. When I left Waunakee, the cable seemed to be all right. Just before La Crosse, the darn thing broke again. For the rest of the trip, I had to set the engine RPM's by the throttle handle position and use my watch to time the flight.

From La Crosse, the route of flight went to Albert Lea, Minnesota, Spencer, Iowa, and then to Sioux

City, Iowa. To avoid landing at the main Sioux City airport, with its control tower, I went to a small airport on the west side of the Missouri river called Martin Field. It has a long narrow asphalt runway with trees on both sides. When I landed, it seemed as if the runway was not much wider than the Cub's landing gear. I mentioned the narrowness of the runway to the old fellow who owned the airport. He said they flew DC-3's out of there on a regular basis, but I'll bet the wheels weren't on the asphalt.

He owned a J-3 that had won an award for the best Cub at Oshkosh that year. It had been restored by some fellows in the local EAA chapter. The deal was that they would fly it for three years after the restoration, then it would be returned to him. Unfortunately, I did not take the time to look at it, but the story seemed plausible as there were several disassembled Cubs in the hangar.

The next leg was to Grand Island, Nebraska. The Cub and I flew the 136 miles in two hours at a ground speed of 68 mph, which is a good speed. This time there was no red carpet to greet me at the ramp.

The final leg for the day was across the desolate prairie to North Platte, Nebraska. By now there were quite a few familiar landmarks along this leg. One of the prettiest is the small village of Eddyville that is located about half way to North Platte and is a few miles off the course line. It sits on the south side of a small river in the bottom of a broad valley. A few cottonwood trees grow on the river's banks. A railroad track goes through town and then disappears into the brown hills. Just northwest of town is an iron truss bridge over both the river and railroad track. This bridge connects a dirt road with the main highway into town. Given the broad expanse of the countryside, Eddyville is a welcome sight in an otherwise endless landscape of brown prairie and wash gullies.

While flying low across this prairie, I am truly alone and the Cub's shadow is my only companion. As I watched the shadow accompanying me, I began to realize that it was a good indicator of my ground speed. If it took quite a while to move from one gully to the next, the ground speed was low. If the time was short, the ground speed was good. Because of this, I started paying attention to the relationship between the computed ground speed and the rate at which the Cub's shadow was moving across the countryside.

The Cub and I finally reached North Platte's Lee Bird field after sunset. We had flown 650 miles in 13 ½ hours on the clock. After refueling and tying down the Cub, I walked to the nearest motel only a half mile away. One of the tricks I had learned was to look for motels while in the airport's traffic pattern. Thus, if I needed to stay overnight, I had a good idea as to how far away the motels were.

Monday dawned with clear skies and a rather strong wind out of the northwest. Leaving North Platte, the course line roughly follows Interstate 80. Using it as my frame of reference, I determined that I had a 45-degree wind correction angle and a slow ground speed. I had a strong sense of de-je-vue as this was the same place that I had encountered strong winds on the Prescott trip. Clearly, The Cub and I were not going to reach Julesberg, Colorado today. Rather than stay on the course line, I followed the Interstate to Ogallala, Nebraska, due west of North Platte, for a gas stop. It took an hour and 15 minutes to go the 55 miles, a ground speed of only 48 mph. It took only 35 minutes to fly the 30 miles from Ogallala to Julesberg and another gas stop. After refueling the Cub, I walked over to the International Harvester building. I told Perry Case about the rest of my trip to Prescott, Arizona and flying back home with Ann. He mentioned the photograph of Rollins Pass that I had sent him earlier.

The route of flight from Julesberg to Sterling, Colorado is along Interstate 76 and the South Platte River. Once, John had flown this leg in his Mooney Mite when the weather went bad. He figured that the best place to fly was over the median strip between the two parts of the highway. Near Sterling, much to his surprise, there was a tall radio tower standing right in the middle of the median strip. So much for that theory. Today, the weather was typical Colorado severe clear. I could see the tower from many miles away.

The airport at Sterling sits on the side of a hill. Landing to the southeast can be a bit interesting as the runway slants downhill in that direction. Today, the wind was from the north, so I landed uphill with a negligible roll out. After refueling the Cub, I took off and took up a southwesterly heading that would take me to the JEFFCO airport--109 miles away. This leg initially goes cross cattle grazing land. About half way to JEFFCO there are three large reservoirs just to the right of course. The South Platte River was crossed at this point. Now, an occasional oil well could be seen in the midst of the ranch land. The line of flight took the Cub across Denver's northern suburbs and then over the city of Broomfield.

The weather held and I finally landed at JEFFCO with its big city traffic and a control tower. The NARCO radio worked well and this time I used the proper radio frequencies. The Cub was tied down to the steel cables that ran along the ground. One winter, when John had the Cub, it survived 80 mph winds while tied to these same cables. However, a wing strut was bent when a tie-down fitting slipped down the strut. After that, I installed tie-down fittings that attached to the wing strut bolts, so they could not slide down the struts.

Tuesday afternoon, John, Joye and I went to the Boulder Airport. There was a high cloud cover and a strong gusty wind. John said that I could take his Mooney Mite, N101C, for a quick trip around the pattern. I had never flown a low-wing airplane or one with a retractable landing gear. In light of the latter, John instructed me to leave the gear down. I climbed into what is basically a miniature fighter plane, taxied out and took off to the west. Once in the air, it was like sitting on a barstool in the sky. I sat rather high in the cockpit and could see downward over the leading edge of the wing. The view was terrific and I was tempted to pretend to be looking for ME-109's.

After about 10 minutes of this revery, it was time to land. The east end of the Boulder runway has about a 50foot drop off that with a west wind causes a strong downdraft. Today there was a 30 degree cross wind from the northwest, not the ideal conditions to land a strange airplane for the first time. I set up my final approach with an extra 50 feet of altitude added for the inevitable downdraft. Sure enough, as I came over the threshold the bottom dropped out. I found myself 10feet off the ground and the Mooney Mite being buffeted by the gusts. I couldn't use the usual Cub technique of a low upwind wing and lots of opposite rudder. Instead, I used the British crab technique where you fly down the runway with the nose pointed into the wind and looking sideways out of the cockpit. At a foot or so off the ground, the power is chopped and a healthy kick on the rudder straightens out the plane a split second before it touches down. This is why the British wreck so many Tiger Moths landing in crosswinds. It worked perfectly for me even though it was the first time I had ever done it. Given the blustery wind and cold, I was very glad to be back on the ground without bending John's pride and joy.

In the summer of 2001, I discovered 101C tied down in the grass at the Portage, Wisconsin airport. John had sold it many years ago and had lost track of who owned it. He was surprised to learn that it was still around although in rather rundown condition. Joye's name was still on the aircraft registration slip in the cockpit.

Chris had joined us, so the next morning (Wednesday) we loaded John's car with bag after bag of rafting and camping gear. The three of us headed west on Interstate 70 for Utah with John's Rogue River dory boat on the trailer behind the car. Driving through the narrow Glenwood Springs Canyon, I looked up at the tops of the canyon walls. Ann and I had flown along those same cliffs looking down at the road we were now traveling. It was quite a contrast in perspectives. After Grand Junction, we drove through the high desert country to the town of Green River, Utah where we camped overnight.

In the morning, a beat up former school bus appeared. Then, the assembled group of rafters threw their gear onboard. Near Vernal, Utah, the bus turned off the highway and lurched its way across the desert, on a narrow bumpy dirt road, to the Green River. Once there, we launched the dory boat and loaded our gear in it. The flotilla consisted of one large raft--a pig boat, several six-man rafts, and our dory boat. The river was rather wide, medium brown in color and slow moving. As the group went downstream, the canyon walls got higher and higher until they matched those of the Grand Canyon in height. After setting up camp that evening, the staff cooked a very nice meal and the tourists became acquainted.

Friday was spent floating lazily down the river and going through a few tame rapids, one of which John swam through. Serving as the oarsman, Chris took the pig boat through several rapids. He also spent time as a paddler on a six-man raft. Going through the rapids in the small rafts can be quite exciting.

Several of the boatmen were college age girls. One of their tricks was to have one of them lay on her back on a raft. Another girl would stand above her and pour a whole can of beer into her mouth from a height of about four feet. A whole can of beer could be drunk this way without spilling a drop–the benefit of a college education.

The flotilla stopped for lunch at the ruins of an old farm on a river flat. While walking around the grounds, I found a Piper Cub landing gear axle. In the distant past, some Cub pilot had landed in the hayfield and the landing gear hit a rock hidden in the grass. I'll bet he had a real problem repairing the airplane at the bottom of a 3,000foot deep canyon 60 miles from the nearest hint of civilization.

In the evening, Mike (the trip leader)announced that since we were on water he was a ship captain and could perform marriages. It was declared that a boatman, nicknamed D-9'er, and one of the young ladies on the trip would get “married” that night. D-9'er always wore a Caterpillar Tractor Company baseball hat, hence the moniker. After diner, a pulpit was set up, John wore a shirt and tie that mysteriously appeared out of his duffle bag. Now, he was suitably dressed to give the "bride" away. About 9:00 P.M.,Mike performed the ceremony. By this time the bride and groom were pretty well under the weather. They were escorted to the sandy beach and treated to a chivaree. Finally, about midnight everybody wandered back to their tents for the night leaving the "newlyweds" on the beach. John and I decided that the real reason the paying customers were on the trip at all was to finance the fun and games of the staff.

After three days of rafting, the flotilla reached our destination--the town of Green River. The rafts were unloaded and deflated. The three of us helped Mike and his staff store all the rafts and gear on trailers. The "bride" bade D-9'er a fond farewell and we all headed back to Denver.

Once there, Chris and I spent a day at John's putting Chris's backpacking gear in a box. Then, the box was shipped home via UPS. I did keep the two-man tent and two sleeping bags just in case we had to spend a night on the prairie.

At 8:00 A.M., on Tuesday the 19'th of August, Chris and I took off from JEFFCO Airport and headed north for Cheyenne, Wyoming. As usual, the weather was very nice. The Cub slowly climbed up to an altitude that would clear the high ground at the north end of the leg. After refueling at Cheyenne, I taxied the Cub to the east end of runway 26. Once cleared for takeoff by the tower, I gave full throttle and the Cub started to roll to the west along the 9,200 foot long runway.

After a very long takeoff roll, the Cub was in the air at an altitude of about 15 feet. Although we were off the ground, the Cub was not gaining altitude. I was a bit surprised as density altitude was not an issue. There was an Interstate highway intersection directly ahead with tall light poles all around it. By doing a few gentle rudder skids and being careful, I was able to fly past the forest of light poles. Once on the other side, our troubles were not over as we were only a few feet above the house tops. I flew the Cub gingerly along a couple of streets and finally reached open country.

After 10 minutes of flying along the edges of hills, the Cub was high enough to allow us to get on course to Scottsbluff, Nebraska. Fortunately, Scottsbluff is some 2,000 feet lower than Cheyenne, so we did not have to do any more climbing. At the end of the first 40 miles, the check point is a microwave tower at the intersection of two power lines. When we got there. Chris's navigation had us right on course. At this point, the ground starts a perceptible slant downhill. Other than the dirt roads there were few signs of civilization and not even a car or truck were seen. About 30 miles past the tower, we saw a lone ranch that was depicted on the map, but no people were seen.

I wanted to take some photographs of the famous bluffs, so I opened up the top of the side door and took the camera out of its case. As I turned to take a picture, the camera case floated out of the door. I made a grab for it, but it went by a few inches from my fingertips. I did, however, take a number of pictures of the famous bluffs. After landing at Scottsbluff, the Cub was taxied across fresh tar to get to the gas pit, which did not please me at all. Pulling the prop through revealed there was no compression in one of the engine's cylinders.

Due to the energy crunch, the Waunakee airport had run out of 80-octane gasoline. Of necessity, I had used 100-octane fuel all summer. The problem is that high-lead 100-octane burns much slower than 80-octane. As a result, the exhaust valves open before the fuel is completely burned Eventually the exhaust valves are damaged. Unfortunately, we were 600 miles from home, had been using considerable 100-octane, and needed to get home before another exhaust valve gave up the ghost.

From Scottsbluff the next leg was to Chadron, Nebraska for a fuel stop. The country is beautiful rolling prairie with a few scattered ranches and not much in the way of navigation checkpoints. Ten miles south of the Chadron airport, the ground rises gently to the north. There are numerous long strings of pine trees following creek beds and gullies. At the north end of this rise is an area of cliffs with light sand colored rock outcrops that contrast with the brown grass and green pine trees. Below and beyond this area, it is prairie again. The Chadron airport is four miles west of town and is just north of a east-west highway.

It is 82 miles, on a northerly heading, from Chadron to Rapid City, South Dakota. It is prairie all the way and there are only a few roads for checkpoints. South of Rapid City there is an airstrip next to Custer State Park, so I tried to talk Chris into camping overnight there. He had been complaining a bit about not feeling well and wanted to push on as far as we could, so we flew on.

After a quick gas stop at Rapid City, we flew a dog leg route to Pierre taking us past the airport at Philip. As the Cub flew over this airport, we could see people on the ground watching us fly by. I thought about landing for gas, but the wire of the gas gauge seemed to be tall enough for there to be enough fuel to reach Pierre some 69 miles away. This turned out to be a bad decision as beyond the point-of-no-return the headwind picked up considerably.

Twenty-five miles before Pierre, the bend in the gas gauge wire hit the top of the tank cap and the Cub was low on fuel. I started keeping an eye out for places to land. The last few miles into the Pierre airport were across the city and then over the tall, but slanted, bluffs on the east side of the Missouri River. As the Cub passed over the city, I had visions of the engine coughing and quitting anytime. In the midst of a city is no place to make a forced landing as what few open spaces that do exist are small and surrounded by trees and buildings. About all that I could do was hope that the engine kept running. After what seemed like an interminable period of time, I landed the Cub, just after a Luscombe, on a runway under construction. I did not feel we had enough fuel to make an approach to the main runway. It was crosswind anyway.

Chris and I watched with great interest as the lineboy filled the Cub's tank. He put in 10.9 of the 12 gallons possible. That was the most gas I had ever put in the Cub at one time on a trip. When we landed, I thought there was enough fuel in Cub's tank for about 15 minutes of flying time. I vowed never to overfly airports with fuel if there was any doubt of getting to the next airport. Many years later, I discovered, that when the Cub is sitting on level ground, its tank holds only 11.2 gallons. At Pierre, we actually could have flown for only another four and one-half minutes on the .3 of a gallon remaining in the tank.

By now it was late evening and Chris and I needed a place to stay for the night. After making a few phone calls, it was determined that there was only one motel room left in town. Since we had the tent and sleeping bags, Chris and I elected to camp, so the couple in the Luscombe could take the room. We rented a car and drove down a long steep hill into town. The campground was in a city park on the Missouri River flats. There were widely spaced cottonwood trees and a number of campsites. Chris and I set up the tent and put the sleeping bags inside. We were the only ones staying in the campground. This was a bit surprising as it was the peak of the tourist season.

Wednesday began with a medium height cloud deck, but it was cold, and breezy. The route of flight was similar to "the northern route" with a gas stop in Huron, South Dakota and then on to Marshall, Minnesota. Arriving at Marshall, Chris and I could see that the only runway was alive with bulldozers ripping up the asphalt and loading it into large dump trucks. One of the problems with the Cub is that if I can't land at my destination there is not enough fuel to fly to an alternative airport. In this part of the country, the nearest airport can be 50-60 miles away. Looking the situation over, I noticed that the trucks were not driving down the taxiway parallel to the runway, so I landed the Cub on the taxiway and then went to the gas pit.

When I went in to tell the lineboy we needed fuel, the fellow behind the desk was quite surprised that we had landed. He said that, due to runway reconstruction, this was the first day the airport was closed. When he learned we were flying a Cub, he realized we didn't land on the runway. Chris and I spent some time telling him about our trip.

From Marshall, we flew to Redwood Falls, Minnesota where there is an FSS. For some reason, getting any information from the man in the FSS was a task for an international diplomat. Then, once the Cub was back in the air, he asked us for a pilot report on the cloud base. Since the Cub was only at about 100 feet at the time, I gave him a wild guess and turned off the radio before he could respond. After Redwood Falls, the Cub followed the Minnesota River to Mankato. Chris and I dropped in unannounced on the folks for the night.

Once in the air, the next morning, our ground speed was quite low. After our experience at Pierre, we stopped at Owatonna, Minnesota for gas. The FBO said that the only aviation gas he could get was the high lead 100 octane. I was a bit annoyed as the Cub already had one cylinder with no compression and really did not need another one to go bad.

For the past two days, when I filled the Cub's gas tank, the cockpit would have a strong smell of gasoline. It would go away after about 20 minutes of flying time. Sitting in the front seat, I did not seem to be bothered by it very much. In the back seat, however, Chris was getting a lot of fumes. Still flying into a strong headwind, we could not make La Crosse nonstop even from Owatonna. As the Cub approached Rochester, we needed to land there for fuel.

By now Chris was not feeling well. I opened up the door and lowered the left side window to give him some fresh air. Being in the back seat, he had to tune the radio to the Rochester tower frequency. By the time he had the radio setup properly, we were looking at the nose of a Lear Jet that just had taken off. I hurriedly called the tower and asked for landing clearance. The tower operator answered in the affirmative, but chided me for being in too close when I first called. I told him that the radio was in back and my operator was airsick. That seemed to satisfy him. After we landed, he let me stay on the tower frequency, which was a nice gesture.

Once on the ground Chris, did not want to get back into the Cub. He wanted to take the Greyhound bus the rest of the way home. After about 45 minutes of fresh air, he decided to fly the rest of the way as long as we kept the windows open. With a quick stop for gas at La Crosse, we were home by mid-afternoon on Thursday.

Fixing the leak

The gas tank in a J-3 was a weak link in its design. The tank was constructed from terneplate, a soft steel sheet coated with a thin layer of lead. The fuel outlet was in the center of the bottom of the tank. When the Cub is parked, the bottom rear of the tank is 1" lower than the outlet. Any water in the tank will collect at the rear and cause rust. I removed the tank and inspected it. No holes were found in the usual places underneath the lower rear tank strap. Finally, I used a needle to pick out some paint in a solder seam about two inches from the top of the front of the tank. Then, I saw a hole so small that the point of the needle would barely go in it. I deduced that after the tank was filled, gas would seep through the pinhole and vaporize. Once the level of the fuel was below the hole, there was no pressure to force fuel through the hole and the vapors would stop. As this was the second terneplate tank to develop leaks, it was replaced with an aluminum gas tank that ended the rusting problems and was fume free.

Platte River at Grand Island Nebraska

Middle Loup River in Nebraska

Hill Northeast of Cheyenne, Wyoming

Scott's Bluff Nebraska

Chapter 8
Another Adventure With Ann

By 1978, it had been five years since Ann and I had taken a long trip in the Cub. Since she and I had been almost to the Atlantic Ocean on our Washington, D.C. trip, the idea of flying the Cub to the West Coast appealed to both of us. After some discussion, Seattle, Washington seemed like a good destination. To make effective use of Ann's week of vacation, we would meet in Billings, Montana. The plan was for the two of us to fly the Cub from Billings to Seattle and then back to Billings. From there, Ann would take the airlines back to Midland, Michigan.

The initial line of flight was to be from Waunakee to Rapid City, South Dakota. Although I had flown this route a number of times before, it had always been from west to east along the "northern route" from Denver. The initial part of the planned route was to be a bit south of the previous route. I wanted to avoid La Crosse as a control tower had been put into operation there.

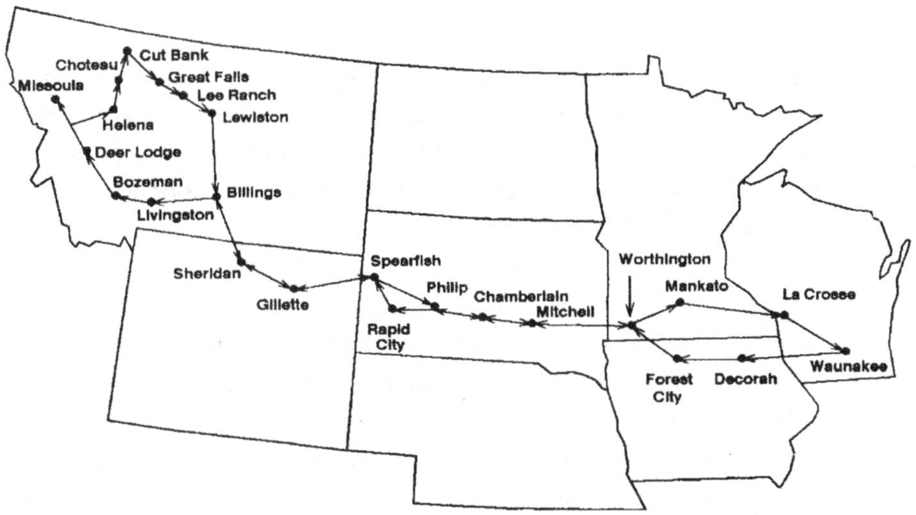

The Route of Flight: Seattle Bound

On my first few cross-countries, I drew the course lines on the maps and also put the compass headings on the maps. Once in the air, I had to get the maps folded, oriented, and read the compass heading. While fiddling with the maps, I often had difficulty getting the Cub properly established on course. To make things easier, I had started the practice of laying out a form on a half of a manila file folder. The names of the cities at the beginning and end of each flight leg, the distances, and the compass headings were written on the form. This greatly simplified things as I could read the sheet before takeoff to get the correct compass heading. Then, in the air, use the map to pick up the course line. Each time I landed, the engine hours' meter reading and the amount of fuel purchased also were recorded on the form. From this information, I could compute the average ground speed and the fuel consumption per hour. This information was used to evaluate my plan for the next leg. The records were also useful for reconstructing the flights at a later date.

The Cub and I departed Waunakee early on Thursday, August 17, 1978, in beautiful flying conditions: clear skies, light winds, and cool for the time of year. My course line took me just north of the Wisconsin River and through the south edge of the un-glaciated area. The countryside was similar to that on the way to La Crosse, but the valleys now ended at the Wisconsin River. The valley of the Kickapoo River contained several small villages. One of them, Soldier's Grove, was flooded so often by this river that the inhabitants received a government grant to move the village to higher ground. A new village hall was built on high ground. Most of the homes were simply picked up and relocated to new lots above the flood plain.

The Mississippi River was crossed at the village of Lynxville, Wisconsin. The river is quite wide at this point and there are many islands for the barge traffic to work their way around. A tugboat was seen pushing a large raft of barges slowly upriver.

A few miles east of Decorah, Iowa there are two beautiful country churches in the midst of the corn fields. The first is a limestone church with a red roof, a tall spire, and a cemetery alongside of it. This church and its cemetery are bounded by rows of tall trees planted in pairs. The second church is a large white painted wooden building that stands out dramatically from the surrounding green corn fields. I used my camera to take pictures of both of them.

I stopped for gas at Decorah and then again at Forest City, Iowa where the Winnebago recreational vehicles are manufactured. The name is painted in red on the roof a large factory building. From there, I flew over corn and soybean fields to Worthington, Minnesota. I was only going to make a quick gas stop here as I wanted to get to Rapid City today. From Worthington it was on to Mitchell, South Dakota. Approaching the city, I could see the famous Corn Palace with its tall yellow ears of corn. The weather was cooperating nicely as from Waunakee to Mitchell an average ground speed of 65 mph was achieved.

The prairies begin in eastern South Dakota. The land is only a thousand feet above sea level, is reasonably flat with gentle rolling features, and is great Cub flying country. In my mind, Chamberlain, deep in the valley of the Missouri River, is where the west begins. To the east of Chamberlain, there is still a fair amount of farmland mixed in with the evolving ranch land. To the west of Chamberlain, the countryside is rolling range land with only an occasional crop field. The vista from the river's east bluffs gives you a very strong sense of what the west looked like when the Indians owned it. A set of undulating hills climb up from the Missouri River bottoms toward the western horizon. These hills look barren, but do have the usual skimpy range grass that is mostly brown by this time of year. A railroad crosses the river on a long low trestle that has two iron truss spans at the east end. These trusses must be there to allow boats to pass under the bridge. A few hundred yards north of the railroad bridge, Interstate 90 crosses the river on a low concrete bridge. This bridge did not have a raised section for boats to go under it. Both the highway and railroad quickly disappeared into the western hills.

The Chamberlain airport is on the valley floor east of the river and north of the Interstate highway. There was not much there, an east-west hard surfaced runway, one hangar building, a house trailer, and a dirt road that ended at the ramp. An old man came out of the house trailer and refueled the Cub.

The next gas stop was at Philip, South Dakota--113 miles away. The line of flight was slightly north of due west. Once past the valley of the Missouri River, the Cub was over the seemingly endless prairie. Regardless of which direction I look the gently rolling prairie has the same dun color of the dry range grass. The only signs of human habitation are the occasional dirt roads with an accompanying line of power or telephone poles. This vast prairie is so lightly populated that I rarely if ever see people. While flying over this desolate countryside, I get a strong sense of detachment from the world. It seems as if I am truly all alone in a domain all of my own. The Cub and I seem to be its only inhabitants other than hawks that occasionally soar silently across my path. After more than an hour of such total isolation, I could finally see Interstate 90. It was the halfway point of the flight segment, and I no longer felt totally alone.

Just past this four lane highway, there is a 350foot tall radio tower on top of a small hill that is exactly on the course line. After an hour of only an occasional dirt road as a check point this tower and the highway were welcome sights. Once past them, the Cub and I were out over the prairie and into our own world again.

Some ten minutes later, I saw a dirt road that ended at a parking lot in front of a white painted wooden country church. The building had a tall pointed belfry and attached to the front of the church was a rectangular entrance that seemed to be an afterthought. This church sat alone in the midst of a vast prairie. There were no other signs of civilization as far as I could see in any direction. This made me wonder where the parishioners lived.

On the way to Philip, I ascertained that my radio was not working very well. I did not need the radio to land at Philip, but would at Rapid City. At Philip, the Cub's tank was topped off. Then, I used the FBO's telephone to call the control tower at Rapid City. I told the tower operator that I was coming in no-radio and gave him an approximate ETA. He was very cooperative and said that he would use the signal lights to clear me for a landing. Thirty two miles west of Philip, the Cub and I flew a few miles north of Wall Drug, a tourist-trap in the middle of "nowhere." It is famous for its signs along the highways all over the United States that list the number of miles from there to Wall Drug. From here, I also could see the tops of some of the rock formations in the Badlands National Park.

When I got close to the Rapid City airport, I called in on the radio. Much to my surprise, the tower operator replied immediately. The radio had apparently cured itself, so there was no problem landing. The weather had been nice all the way from Madison to Rapid City where I landed at 7:30 P.M.. The 683 miles were flown in 10.36 hours of flying time at an average ground speed of 68 mph. This was a great deal above my lifetime average of 55 mph.

The telephone bank in the airline terminal was used to call a number of motels without finding a room. After about 10 minutes of this fruitless activity, the young lady at the Avis car rental desk noticed my plight. She found a room for me at the Hiram Johnson hotel. I was very grateful for her assistance. The hotel dates from the days of the old west. It has a marvelous lobby dominated by a huge stone fireplace. The wall above it is decorated with deer and buffalo heads. The management was quite proud of the fact that Teddy Roosevelt had been a frequent guest.

The next morning(Friday), the weather was very nice again with partly cloudy skies, not much wind, and rather cool. The destination for the day was Fort Smith, Montana where I planned to camp overnight. This path took me into unfamiliar country. On previous flights, I had come from the south and turned east at Rapid City. Now, I was going to turn to the north.

After takeoff and saying goodbye to the control tower, the Cub and I entered a valley whose center was occupied by Interstate 90. The valley went to the northwest along the eastern slopes of the Black Hills to Spearfish, South Dakota where I landed for gas. Strangely, the local airport denizens were definitely unfriendly. From a newspaper clipping on the wall, I learned that the airport had been founded in the 1920's by a aviation pioneer named Clyde Ice. He was from Waunakee--small world.

After takeoff, the Cub was flown around the north end of the Black Hills. They were covered with pine trees that were a very dark shade of green. There also were numerous light grey rock outcrops accenting the dark coloring of the hills. Once around the north end of the Black Hills, the course went to the southwest, then around the south end of a small set of high hills, and then due west into open country.

Near Gillette, Wyoming, long lines of railroad cars were waiting to haul coal to the Midwestern power plants. There also were several large buildings associated with the underground coal mines. I didn't know that the Gillette airport had recently acquired a control tower, so I just came in and landed--a practice the FAA frowns upon. After parking, I saw a portable control tower mounted on a small blue ex-Air Force truck. I gave the tower operator a call on the radio and apologized for the unannounced landing. He said that most of the ranchers flying into Gillette don't have radios, so he was used to such arrivals. He hadn't even noticed me until I taxied by the truck.

The flight segment from Gillette to Sheridan, Wyoming lies parallel to the east face of the Big Horn mountains, but over the front edge of the foothills. Just north of Gillette, there is some unusual countryside. Every few hundred yards there were natural mounds about 5060 feet tall, shaped like small volcanos, formed by erosion many eons ago.

All along the route, there were numerous broad flat-topped ridges. In this part of the country, they call them buttes. The sides of the buttes were steep cliffs and small creeks were in the valleys below. The tops of the buttes were covered with range grass that had a distinctly green appearance. It was not the lush green of Midwestern fields, but a more subdued green that often faded to a light tan color in the distance.

The tops of many buttes had dirt roads that were used by the ranchers to visit the cattle grazing there. Even though it was very desolate, I felt quite comfortable flying here. I could land the Cub on most of the flat areas, but would have to be on the lookout for rocks hidden in the grass. Occasionally, antelope could be seen grazing in the midst of a group of cattle. That seemed to be a very good idea to me. The segment was only 77 miles long and the flight was uneventful.

The airport at Sheridan has a long north-south runway and a cluster of buildings on the east side. In the center of them was a small 1930's white stucco building with a long abandoned control tower capping its roof. After refueling the Cub, I asked one of the lineboys about camping at Fort Smith. He and several other locals indicated quite forcefully that it was not safe to park an airplane overnight at the Fort Smith airstrip. The airstrip is located on an Indian reservation and the campground was on the other side of the river. Looting of unattended aircraft was prevalent. It was not considered safe to camp at the airport. Consequently, I planned for a nonstop flight to Billings, Montana--93 miles away.

Once in the air, I located a power line that paralleled my course from Sheridan to the Yellowtail dam and kept it within sight. The dam is an impressive sight as it is across the east end of a deep canyon in the face of the mountains. The dam is very tall with a reservoir behind it that winds through the mountains for many miles. From my low altitude, I could see the dam and only a short stretch of water behind it. The Big Horn River flows out of the base of the dam and goes to the northeast on its way to joining the Yellowstone River.

The Fort Smith airport was on the north side of the river. There was a lone Cessna 172 parked alongside the single asphalt runway. Reluctantly, I overflew the airport and headed toward Billings now only 45 miles away. This leg was quite scenic as I was flying over ranch country interspersed with higher flat topped buttes. Flying from the back seat of the Cub, I had a wide angle view of a landscape in which the Big Horn mountains were receding to the west, the buttes were in front of me, and to the east was the sunlit rolling prairie. Thus, the details of this panorama slowly changed as the Cub and I moved along the route of flight.

The city of Billings is on the valley floor north of the Yellowstone River and is nestled against the base of a tall butte. The airport is situated on top of this butte and has a large airline terminal. Northwest Airlines has served the city since the 1920's.. Several of their Boeing 747's could be seen parked on the ramp. Clearly, the airport was an unfamiliar environment for the Cub. The airport has a control tower, but I had no difficulty making contact via the Cub's radio and getting landing clearance.

The good weather along the route had allowed the Cub and I to arrive a day earlier than expected. It had taken roughly seventeen hours of flying time to cover the 990 miles from Madison to Billings yielding an average ground speed of 59 mph. When the Cub landed at Billings, an important milestone was reached—its engine hours meter read 1,000.39 hours.

The next day (Saturday, August 19), Ann's flight arrived midmorning. When she came out of the terminal, I was appalled at how much baggage she had brought along. The Cub has very little storage space and is placarded to carrying no more than 20 pounds of baggage. I already had the tent, two sleeping bags, a cook stove, my clothes, and food for three days stuffed in the baggage area. After much pushing and rearranging, we finally crammed all of her gear into the plane. Due to the nature of the terrain we were going to fly over, I sat in the front seat to serve as pilot while Ann, in the rear seat, acted as navigator. Once organized in the cockpit, I gave the tower a call informing them that we were ready to go. They cleared us to the active runway. With a bit of throttle, the heavily laden Cub began waddling along the taxiway to the east. The runway slopes up to the west and ends at the top of a tall cliff.

Once in position on the runway, the tower cleared us for takeoff. I pushed the throttle to full power and the Cub started to roll. After going an inordinate distance, the Cub finally lifted off the runway, climbing to a height of about six feet. Much to our collective chagrin, the Cub would climb no higher. The end of the runway was rapidly approaching and beyond it was the cliff. If we flew off the cliff, there was no way the Cub could climb back up to the altitude of the airport. I quickly pulled back the throttle, landed, and called the tower:

"Billings tower, 88354 is aborting the takeoff."

"Do you want to declare an emergency?"

"Negative, we are just over gross."

"Taxi back to the ramp."

"Roger."

As Ann and I taxied the Cub back to the ramp, I passed close behind one of the 747's to take a picture of it. The tower warned me that I was too close to it. I thanked the tower for the warning and taxied a bit further out.

Back at the fixed base operator's facility, Ann and I pulled everything out of the plane. Then, we decided what was essential and what was not. Our surplus items were placed inside a cupboard in the shop. I told a mechanic that we would be back for our belongings in about a week.

The second takeoff was slow, but uneventful. When a reasonable altitude was reached, The Cub was pointed toward Livingston, Montana some 93 miles west of Billings. As we flew to the west, the skies became overcast. Halfway there, it started raining lightly, but to the south there was a high broken cloud layer with the sun peeking through it. Livingston is at the end of an amphitheater shaped valley where the butte country meets the mountains. The city was hemmed in on the south and west sides by mountains. Because of this and the clouds, the further we flew into the valley the darker it became. I had the distinct feeling that the sun did not shine on the town very often.

A few miles before Livingston, the Cub encountered a very strong downdraft caused by a west wind spilling over the mountains and down into the valley. The Cub started to sink at an alarming rate. I gave full power and pulled back on the stick to try to arrest the descent, but this had no effect what-so-ever. The old bird just kept on sinking. As the Cub started to get dangerously low, I looked around for a spot to land, but the ground was strewn with boulders. A landing would surely mangle the Cub. I had read once that downdrafts tend to dissipate when they get close to the ground. It was obvious that Ann and I were going to find out if the author knew what he was talking about.

On a good day, the Cub is hard pressed to climb at 200 feet-per-minute. Thus, trying to climb up in a strong downdraft was a lost cause. At best, climbing would soften the impact when we hit the ground. It seemed prudent to try to fly horizontally out of the downdraft. I leveled out and gave the engine full throttle. As usual, Ann sat in the back, cool as a cucumber, watching the proceedings and offering a few running comments. All the time the Cub was sinking closer and closer to the ground. Finally, at about 30 feet above the ground, the downdraft dissipated. In a few seconds, the Cub was clear of the downdraft and Ann and I breathed a collective sigh of relief. This was our first encounter with a prolonged downdraft. We were not too thrilled about the prospect of meeting another.

At the Livingston airport, the Cub was refueled. Then, Ann and I checked the weather at the FSS on the airport. Other than the solid overcast and cool temperatures, it was good Cub weather. By now it was late afternoon and the dim light was giving the valley a spooky appearance. Both of us were not about to spend any more time here than necessary. Between Livingston and Bozeman, Montana there is a 6,002foot mountain pass that is 1,346 feet higher than the valley floor here. The Cub, Ann, and I spent the better part of 45 minutes flying back and forth over the hills looking for some lifting air. The heavily weighted Cub just didn't like going high enough to clear the pass.

From our vantage point, Ann and I could see where the Yellowstone River went through a short gorge then out into another mountain valley. Although our valley was dark and gloomy, the floor of the other valley was lit by rays of golden sunshine streaming to the ground from holes in the dark clouds. The sight of the granite mountains, the green fields, and the lit and unlit areas on the valley floor made a memorable scene. One of my university colleagues and fellow skier has a retirement home on an island in the river near where it leaves the sunlit valley. As we passed the mouth of the gorge, I took a photograph of the river valley and later gave him a copy of it.

Finally, after flying along every hillside in this end of the valley, enough altitude was gained to be a few hundred feet above the pass. Then, the Cub was pointed west to go over the pass. Just over the center of the pass, a Cessna 172 flew right underneath us. They were so close I could see the two people in the cockpit through the Cessna's overhead windows. Given the Cub's propensity for losing altitude, for no apparent reason, that did not seem like a particularly smart maneuver on their part. Once over the pass the Cub entered a broad mountain-ringed basin.

Even though Northwest Airlines served Bozeman, there was no control tower. I simply made a long straight-in approach to the west and landed. After tying down the Cub, Ann and I went into the Bozeman airline terminal. It is constructed out of local stone and huge rough-hewn timbers. The interior has a high vaulted ceiling. Descending from the vertex is a set of large geese fabricated from copper sheets. Each bird is in a different phase of the landing process and the overall effect is stunning.

I had planned ahead for this first night and had made a reservation at the local Holiday Inn. After checking in, Ann and I spent the evening walking around Bozeman and recalling what we had done here many years earlier on a family camping trip. In particular, Ann remembered the many pools in the hot springs.

The next morning (Sunday), the Holiday Inn's van took us back to the airport. Ann and I loaded up the Cub for another day of flying. It was on to Butte, Montana with a somewhat higher pass before us. Since it was early in the morning and rather cool, the Cub had no trouble climbing up to the desired altitude. I was surprised that the floor of the mountain basin was covered with large wheat and alfalfa fields.

Interstate 90 goes from Bozeman to Butte. All Ann and I had to do was follow it. Halfway to the pass, we flew just north of a general aviation airport with many airplanes parked on the ramp. By the time the Cub was close to Butte Pass, the Interstate highway started going up hill. To me, it appeared as if we could not follow the Interstate through Butte Pass. Hence, I deviated a few miles to the southwest to the 6,453foot Pipestone Pass. The east side of this pass slopes up rather gently, but the west side is nearly a cliff. By the time we reached the pass, the Cub was at an altitude of 7,000feet and well above the pass. Squandering this altitude by landing at Butte, 1,500 feet below, seemed to be a waste of altitude. A check of the map showed that the Deer Lodge airport was another thirty-eight miles ahead. Ann and I opted to land there. This line of flight took us within sight of the huge open pit copper mines that had been shut down for economic reasons. Ann and I marveled at their size and multi-hued colors.

Before leaving home, I had looked up Deer Lodge in several airport directories. In all of them, it was listed as being attended and having fuel. After landing there, it was clear that the airport was unattended. Worst of all, there was no telephone by the fuel pump to call for service. Getting fuel at an unattended airport is a time-consuming activity that can cost the better part of a day.

Ann spotted a National Guard facility on the west side of the airport where a helicopter was warming up. The two of us ran across the fields to the chopper. I explained our situation to a master sergeant standing at the chopper's door. The crew was about to depart to a county fair where they were to spend the day on public relations duty. He told the pilot to go on without him and that he would stay to help us. He went inside the building and came back with several five-gallon Jerry cans. He placed them in the General's jeep and drove us into town. After filling the cans with automotive gas, we returned to the airport.

The Cub was topped off and the rest was poured into the jeep's tank. As I recall, the master sergeant's name was Smith. He was in his early sixties and had the straightest ram rod posture I have ever seen. It was gracious of him to forgo a day with his crew to help strangers, but at small town airports helping others is the norm.

The mountain valley going to Missoula, Montana begins just beyond Deer Lodge. Both sides of the valley consist of a series of ever higher hills that eventually become mountains. Through a slot between the mountains, Ann and I could see the Bridger Bowl ski area where my Livingston colleague skis.

When the mountain tops are shrouded in clouds, a common practice among local pilots is to get to Missoula by flying through this valley under the clouds. The aviation maps show this valley as an approved low altitude route through the mountains. Ann and I were a bit apprehensive about flying too low through the valley as power lines crossed it periodically. Although each set of wires had big red balls on the top wire, they would be tough to see in marginal weather.

The evergreen trees on many of the hillsides were being cut for lumber. Some hills had been stripped bare, via clear-cutting, giving them a devastated appearance. I'm sure that it would take 100 years for any semblance of forest to grow back again in the poor soil. A few miles past Deer Lodge, we flew over a large lumber mill along side the small river that flows through the valley. There were several large mounds of sawdust alongside the mill, but no piles of logs.

About halfway to Missoula, at the village of Drummond, the valley widened to the south to form a small mountain basin. In a few miles the valley narrowed again. At one point the valley made a sharp turn to the north and then went to northwest again. Today, only medium height clouds were above us, so we flew uneventfully through the valley.

In the last 20 miles before Missoula, the valley became quite narrow and there were tall mountains on each side. Sitting in the front seat, I could look up at the mountains, but could not see their tops. As the Cub approached Missoula, a ridge could be seen that seemed to block the end of the valley. The Interstate made an S turn as it went through a slot in the ridge. Rather than trying to follow the Interstate, I climbed the Cub high enough to clear the ridge and then we could see Missoula just on the other side of the ridge.

Missoula had a control tower, one very long runway and a shorter one crossing it at a slight angle. I called in, landed on the long runway, and taxied to the fixed base operation. Then, Ann and I tied the Cub down on the ramp. It now was 2:00 P.M.

After determining there was bad weather ahead, we walked to the airline terminal in search of a rental car. The Hertz counter girl had just left and no car was to be had. As a result, Ann and I hiked two miles to the nearest rundown motel that was on a par with those in Gothenburg and Blandling. As the motel was on the northern outskirts of the city, we never did get to see the sights of Missoula.

Spokane, Washington was the next major stop on our route of flight. To get there we had to fly over Mullan Pass that was at 5,260feet. The Cub did not have enough range to get to Spokane, so I planned to get gas at the Kellogg, Idaho airport, just on the other side of the pass. Ann and I spent the better part of Monday morning, in Missoula, waiting for the weather over Mullan Pass to improve. All this time, Spokane was reporting low ceilings, rain, fog, and the airport was IFR. Not wanting to get trapped on the western side of the pass, the Seattle destination was reluctantly declared a lost cause. Ann and I decided to go to Glacier Park via Helena and Cut Bank, Montana. I thought it would be interesting to see Glacier Park from the air. Our family had camped there several times while on vacation trips through the west.

After the Cub was pre-flighted, I contacted the Missoula tower, taxied to the active runway, took off to the south and climbed over the ridge at the end of the valley. Once over it, Ann and I headed the Cub back down the valley toward Butte. Incongruously, the weather was perfect: blue skies; high, wispy, cirrus clouds; and little wind. It was hard to believe that Spokane was still IFR.

After flying in the valley for roughly 45 minutes, the Cub reached the village of Garrison, Montana. We made a left hand turn and started to follow Highway 12 to the east. This time the west side of the mountains had the up slope and the east side had very steep drop-off. Flying up the western slope was a challenge. The Cub had to climb some 2,000feet in only 25 miles. The Cub was equal to the task and climbed easily to the top of the 6,325 foot McDonald Pass. As Highway 12 went over the pass, there was a scenic overlook with many cars and camper trailers parked in it. Once at the pass, I had to set up a rather rapid descent to get down to the 4,265foot level of the Helena airport only 15 miles away.

After a brief gas stop in Helena, it was off to Choteau, Montana. The countryside, just outside of Helena, was very different from what we had been flying over. There was a rather flat plain for the first few miles ending in a region with medium sized rounded hills that looked as if they were solid rock. In the midst of these bare rocks was a small lake. Once past the rocks, we were back in rolling country. The Rocky Mountains could be seen on the western horizon.

At the Choteau airport, Ann and I were greeted by a wheelchair pilot in his early sixties who owned an Ercoupe, an airplane that can be flown without using your feet. I asked him about camping at the Browning airstrip near Glacier Park, but he advised against it for the same reasons that Fort Smith was avoided. As an aside, he added:

"The Indians are still looking for me."

Back in the 1950's, he was barnstorming, in an old Travelair biplane, at county fairs and other such gatherings. At a Browning Indian Pow-Wow, an Indian fellow climbed in for a ride. After flying a while he became airsick and indicated that he wanted out right now. The pilot landed in a field, let him out, and then flew away in great haste. What the Indian didn't know, but the pilot did, was that it was about twenty miles of hiking through rough country back to the Pow-Wow. Needless to say, the barnstormer made a point of avoiding the Browning airstrip after the incident.

He drove us to a KOA campground just below the south end of the runway. After pitching our tent in the midst of an empty campground, Ann and I walked the few blocks into town. In the center of town was a rustic log cabin restaurant. During dinner we felt as though we were really in the old west. When Ann and I returned to the KOA campground, we were astonished to see every campsite filled with tourist's motor homes and camper trailers.

The next morning (Tuesday) Ann and I arose early and our breakfast consisted of orange juice and rolls purchased in town the night before. Then, we broke camp, packed our bags, and walked up the hillside to the airport. As we were loading up the Cub, a disappointed Ercoupe pilot arrived. He had driven into the KOA campground to give us a ride back to the airport, but we had already left.

The flight from Choteau to Cut Bank, Montana, under dark grey overcast skies, was within sight of the Rocky Mountains. Despite the clouds, the air was quite smooth at our low altitude. The mountain peaks here were very jagged, tightly packed along the horizon, and they appeared quite ominous. These mountains were more like the Swiss Alps than the Colorado Rockies.

In WW II, the airport at Cut Bank had been an Air Force B24 base, but only one runway and a single large wooden hangar (a duplicate of that at Pierre) remained from those days. The airport was manned by a fellow, who eked out a living repairing airplanes, selling gas, and flying tourists over Glacier Park. We asked him about flying to Kalispell, Idaho. He informed us that we could get there by flying through a series of narrow mountain valleys. I looked at the map and then out the window. The clouds were getting lower and darker making for a very gloomy day. Getting to Kalispell via that route didn't seem to be a good idea.

Instead, Ann and I decided to fly over St. Mary's lake, in Glacier Park, to look at the area from the air. As the Cub approached the park, the clouds above us became multilayered with the lowest layer consisting of wispy strands of clouds undulated up and down in long waves. These strands appeared black against the lighter, higher clouds and were simultaneously unusual and very striking. The cloud patterns indicated the air was very unstable, hence, the weather was going to get worse very soon. The Cub was circled over the edge of the lake for twenty minutes while Ann took pictures. From the air, we could see the top of Sieah Pass that our family had climbed many years before.

Since the Park Service frowns on airplanes flying around at low altitudes, we headed back to Cut Bank before the Cub drew their attention. On one of my Denver trips, I had met a fellow and his girlfriend who were barnstorming, in a Stearman, to earn enough money to fly to the east coast. They had flown low across Yellowstone Park and some park ranger had given their N number to the FAA. That clearly discourages one from hanging around a national park for very long.

The night before, Ann and I had re-planned our flight to go from Cut Bank to Great Falls to Lewistown and then south to Billings. The leg from Cut Bank to Great Falls took us over rolling country devoted to dry land farming. The fields below were alternately plowed earth and ripe wheat giving the countryside a yellow and black striped appearance. Each field is about 300 yards wide, a mile or more long, and flows up and down with the rolling countryside. From our low altitude, the stripes seemed to stretch out to the horizon. The widely scattered ranch houses were painted in bright colors: reds, blues, yellows, and greens. They stood like beacons in the midst of the fields.

About halfway to Great Falls, there was a picturesque railroad trestle roughly a mile off the course. Being a point A to point B person and always nervous about the Cub's limited range, I rarely make side trips off the planned route of flight. However, this trestle was too appealing to be simply observed at a distance, so a detour was made to look at it. The trestle went across a wide, but not very deep valley. The supports were pairs of steel pillars with crisscross members joining the pillars. There were quite a few of these pillar pairs and they appeared to be painted black. The trestle must have been used quite often as the rails were shiny.

A few miles before Great Falls, I climbed the Cub up to traffic pattern altitude and called the tower for landing clearance. It was only 93 miles from Cut Bank to Great Falls, but it took two hours and 13 minutes at a ground speed of only 42 mph. After the last several airports, the Great Falls airport seemed quite large. In the 1940's, Great Falls was a major stop on Northwest Airline's transcontinental route to Seattle. The large white stucco art-deco airline terminal building, with a control tower on its roof, was a remnant of an earlier day. If it weren't for the National Guard jets parked on the ramp, the scene could have been straight out of a 1940's movie.

Despite our lousy ground speed, the weather had improved since we left Cut Bank. There were a few puffy clouds off to the west and to the east the skies were clear. Ann knew what kind of weather we had left behind at Cut Bank and insisted that I check the weather at the Flight Service Station. The briefer indicated light winds held along our route of flight to Stanford, Montana, with thundershowers expected in Great Falls around 7:00 P.M., and to the east it was quiet.

It was 3:00 P.M. when we took off and headed southeast for the town of Stanford (only 57 miles away) in the Great Basin of Montana. Before the turn of the 20'th century this hundred-mile long, mountain-ringed basin was the center of Montana's cattle grazing industry. It was here where the old west lived out its final days before succumbing to "progress."

For me, this part of the journey was filled with nostalgia. In the 1890's, the cowboy artist, Charles Russell, had sketched and painted in this locale while tending cattle. My mother, as a young woman, had lived in Helena and had seen several of his works in their local settings before they found their way into art galleries. In addition, she had a biography of Russell at home. I had read it many times and had practically memorized the paintings in the book. As we flew across the Great Basin, I was mentally matching up the countryside with Russell's art and tried to visualize what it was like when he lived here. Our destination, Stanford, was a town intimately associated with Russell in his cowboy days.

About ten minutes out of Great Falls, my reverie was interrupted by a tap on my shoulder. Ann said "I don't think we have any ground speed." To verify this, she started timing as we went across a road. At the next checkpoint, she calculated a ground speed of only 12 miles per hour. We were flying into a 60 MPH headwind. Ann looked back toward Great Falls and all she could see was a black wall of thunderstorms--there was no turning back. My first thought was to kill the weather briefer if I ever saw him again.

When encountering high winds in the Cub, the first tactic is to fly lower as the wind tends to get attenuated by ground features. Dropping down to about fifty feet, the Cub was immediately tossed around by the turbulence resulting from the strong winds passing over a series of boulder strewn gullies. Repeatedly, the Cub would get pushed down to within a few feet of the ground. When we went down, I would push the throttle to full power, pull back a bit on the stick, and then we would be tossed back up in the air again. Once at an acceptable height, the throttle was pulled back, and the Cub put in its level flight position. Then the next gust would hit and a wing would go up and I had to fight the controls to get level before being pushed down again. All through this I was very busy churning the controls while Ann was viewing the ground beneath us with some alarm.

After what seemed a very long time, the Cub moved out over smoother fields and the ground induced turbulence diminished. After a while, my navigator informed me that our ground speed had improved to 40 mph and that we had enough gas to reach Stanford some 40 miles ahead. While poking along, there was plenty of time to observe the countryside that was devoted to cattle grazing and alfalfa fields. Every mile or so sat a rectangular stack of hay bales roughly 50 feet long, 25 feet wide, and 12 feet high. Each haystack was surrounded by a barbwire fence to keep the cattle from eating the hay. I made a mental note that a haystack would make a good wind break if we found it necessary to land.

The Cub was, more or less, following the highway to Stanford. Some four miles ahead, it went up a steep grade and then over a butte whose top was roughly 300 feet higher than the floor of the basin. The wind had picked up again and reaching Stanford seemed problematical. I told Ann to reach into the baggage compartment and take out the tie down stakes and ropes. There were about 40 pounds of camping gear and luggage on top of the baggage compartment lid. Ann undid her seat belt, turned halfway around, and began the difficult task of pulling out the bags. The only place she could put them was on the seat or the floor. Placing anything on the floor was a bit dangerous as the items could interfere with the movement of the controls.

In the midst of this process, one of the circuit breakers on the instrument panel popped indicating an electrical short circuit. My first thought was that we had taken a lightning strike, but everything looked all right. I reset the circuit breaker and it popped again--to heck with it. In addition, the intercom had quit working. Ann and I were communicating by yelling to each other. Finally, the tie down equipment had been retrieved, the bags were back on the shelf and Ann had her seat belt secured.

I picked a path toward the butte about 75 yards to the left of a deep wash gully that went down the face of the butte. As the Cub approached the lip of the butte, a strong wind was flowing down the lip and the Cub was not climbing fast enough to get over the top. By now, we were quite close to the face of the butte and there was not much working room remaining. I yelled to Ann:

" We are going to turn around and land in a field."

I initiated a 180 degree downwind turn. At roughly 160 degrees through the turn, the Cub's tail suddenly went up in the air as if lifted by some giant unseen hand. Now, we were about 60 feet above the ground with the fuselage in what seemed to be a near vertical nose down position. For a moment, it seemed as if we were suspended motionless in midair. I had visions of the Cub going over on its back. Instinctively, I knew that unless we obtained some airspeed quickly the Cub was going to fall out of the sky right on its nose.

Despite the lack of altitude, I deliberately pushed the stick full forward to put the Cub into a dive. The normal reaction would have been to yank back on the stick in an attempt to pull out, but the one thing we desperately needed was airspeed. The only way to get it was to dive. As the stick reached the full forward position, It felt like we were no longer motionless and were beginning to drop. Immediately, I started to pull back gently on the stick. At the same time, the wash gully, that had been off to our right, now was directly in front of us. At about 30 feet above the hillside and still in a steep dive, there were not a lot of options open to us.

The angle of the slope and the depth of the gully provided a chance for survival. The curved arc of our pullout path put the Cub right into the gully. Fortunately, the gully was deep enough to contain the Cub and wide enough to have about 10 feet to spare at each wing tip. Once inside it, the Cub's wings were even with the top of the gully's banks. Due to the 60 mph tail wind, we were now flying through the gully at a ground speed of around 130 mph. I had no idea of what lay ahead inside the gully and hoped there were no fences or wires across it.

After about 20 seconds, I caught a glimpse of the top of a haystack as we went past it. Immediately, I pushed the throttle to full power, pulled back on the stick, and did a climbing chandelle turn to the right. This took us up and out of the gully. As the Cub turned into the wind, the ground speed rapidly diminished. I rolled the wings level and aimed the Cub at the haystack, which made a splendid windbreak. A full stall landing was made with the Cub touching down about 50 feet from the haystack. The Cub didn't roll more than ten feet. I told Ann:

"As soon as we were close enough, tie the wing struts to the fence."

Ann had done this before at La Crosse. As soon as I taxied the Cub up to the fence, she opened the side doors, jumped out, and went into action. Within a minute, she had tied the right wing strut to a fence post, screwed an anchor into the ground, and tied the tail down. Under the left wing she could not get the anchor to penetrate the ground. The plane also was not lined up with the fence and the third rope was not long enough to reach a fence post. Now that the plane was protected from the wind by the haystack and all but the left wing well secured, I turned off the engine and climbed out. Ann's first words were:

"Gee dad. You are a great pilot".

Just then, a farm truck came over a rise to the east and stopped near us. A slightly built man about forty, in dirty overhauls, and wearing a baseball hat climbed out with a look of total disbelief on his face. His first words were:

" The last time I saw your airplane it was in a nose dive and had disappeared into the ground. I came over to pick up the bodies."

I quickly explained what had happened and our immediate need was to tie down the left wing. He climbed into the truck, drove off, and in a few minutes returned with a big stake and a sledge. He pounded the stake into the ground and I tied the left wing to it. In the meantime, Ann had taken the tent out of the airplane and set it up under the left wing. Overall, we hadn't been on the ground 10 minutes, yet, the Cub was tied down and camp was set up.

A bit later, the man returned with the rancher, his last name was Lee, in whose field we had landed. His CB handle was "Buffalo Bill." The first man was the hired hand, who was originally from Alabama. He had been stationed in the Air Force at Great Falls, but when his tour of duty was over he had stayed in Montana.

In the course of the conversation, they mentioned that a husband and wife had been killed, in a Cessna 172, the previous year at the same place where we had been flipped vertical. The rancher invited us to dinner, but we declined the offer, wanting to stay with the airplane. Just about the time I was regretting leaving all our food in the cupboard at Billings, the rancher returned with several tuna fish sandwiches, a thermos of coffee, and a dozen Oreo cookies that were greatly appreciated.

To the west of us, the line of thunderstorms appeared ominously close with the frontal roll cloud only a half mile or so away. It looked as if the storm would overtake us momentarily. Ann and I anticipated spending the night out in the rain trying to keep the Cub from blowing away. The rancher indicated that as night fell the wind would shift to come from the west at 50 mph. The haystack protected us from a south wind, but not the west. To break the wind, he had the hired hand bring out a large stake truck and park it to the west of the Cub.

That night lying in the tent, I asked Ann what she thought about our flying down the gully. Her response was "what gully?" When we pitched straight down, her mind blanked. She had seen nothing until the wheels touched the ground announcing a safe landing.

Her response reminded me of an incident that occurred during the Korean War at our air base, K-14. One night a shot-up B-26 made an emergency landing. In this version of the B-26, the bombardier sits in the nose and the navigator in the cockpit's right-hand seat. A short narrow tunnel connects the two spaces. Before the navigator can occupy the right-hand seat, the bombardier must climb into the cockpit first and squirm through the tunnel into the nose.

After the plane stopped, the emergency rescue people could not find the bombardier. The glass nose was still intact and the pilot and navigator didn't have a scratch on them. After 10 minutes of pondering the fate of the bombardier, the rescue team decided to take the rest of the crew to the base hospital for a checkup. The medics opened the rear doors of the ambulance and there sat the missing bombardier. After the B-26 had slid to a stop, the bombardier came through the tunnel, reached up and opened the cockpit hatch over the navigator's head, stepped on his seat, and climbed out of the airplane. The navigator had absolutely no recollection of seeing the bombardier get out. So Ann wasn't alone in blanking out in an emergency.

My own reaction to the day's incident was quite unusual, for me at least. During the course of events, I was so busy flying the Cub that all of my thoughts and instinctive actions were focused on getting us (the Cub, Ann and myself) out of a life threatening situation. In addition, if we did crash there was little chance of survival. It was not worth thinking about those prospects. Once on the ground, I was basically unruffled. Close calls usually leave me with shaking hands and wobbly knees. I suspect it all happened so quickly (the whole sequence did not take more than 60 seconds) the danger did not have time to register at an emotional level.

Early in my flying days, I had taken about ten hours of aerobatics instruction with Jake Bloyer, for whom an airplane is an extension of his finger tips. From him, I learned to control an aircraft in unusual attitudes. I suspect this training was the key to my doing the right things when it counted. After returning home, I related the events to Jake. I asked him about my decision to push full forward stick rather than try to pull back. He indicated that I had done the right thing even though it was counterintuitive. Needless-to-say, I thanked him for the aerobatics lessons that gave me the presence of mind to instinctively do the right thing.

The next morning (Wednesday) dawned with a dead calm and a high thin cirrus cloud layer that the rising sun turned into beautiful pink lace against the dark blue sky. I spent 10 minutes just savoring the fact of another day.

Ann and I took down the tent and packed the camping gear in the Cub. I was very glad that we had not left this equipment in Billings. Just as we were getting ready to leave, the hired man showed up to watch the Cub takeoff. The field was a nearly solid mass of small, smooth rocks about the size of a man's fist with the alfalfa plants growing in the small spaces between the rocks. How they harvested the crop without ruining their equipment, I'll never know.

On the takeoff roll, across the short dimension of the field, the Cub kept bouncing around on the rocks. It was not possible to get enough speed, so I aborted the takeoff. Then I taxied down to use the long dimension of the field that was less desirable since the takeoff would be slightly uphill. After much bouncing the Cub finally lifted off and a few circles were needed to obtain some altitude. As we circled, Ann took several pictures of the weather-worn ranch buildings. On returning home, I sent "Buffalo Bill" a framed enlargement of one of the pictures.

Ann and I never did land at Stanford, choosing instead to fly on to Lewistown, Montana (not to be confused with Livingston near Billings). A thin, grey haired woman in her late 60's ran a small, neat restaurant at the Lewistown airport. Her pancakes were a welcome treat. The previous night's storms made the local paper's headlines. Golf ball sized hail had fallen at Great Falls. At Havre, Montana a commuter airliner, on the ground, was damaged by wind and hail. It was a good thing we left Great Falls when we did as hail of this size would have punched large holes through the Cub's fabric covering. Our luck also held in that the storms moved on a southwest to northeast track. Our campsite in the field was just on the eastern edge of this storm track.

The 110-mile route of flight from Lewistown to Billings took the Cub through the Judith Gap that had played such a prominent role in Charles Russell's cowboy days. This gap was the southern entry for the annual cattle drives into the Great Basin of Montana. It is a wide, east-west saddle in a mountain range. The ground also rises from both the north and south to meet at the low point of the saddle. Just as we passed through the gap, the bottom of our seats seemed to drop away and another downdraft had the Cub in its grasp. This one was short lived and was more of a shock to us than a problem.

The weather continued to be beautiful: clear skies and fifty-mile visibility. En route, our ground speed was a less than speedy 53 mph, but it was better than the 42 mph from Cut Bank to Great Falls. After touchdown in Billings, Ann and I felt like we were back in civilization again. Our surplus gear was retrieved from the shop's cupboard, a large cardboard box was found, and the lot shipped to Madison via Northwest Airlines airfreight. Something the barnstormers couldn't do. Not having completed a large part of our planned trip and with a few days of vacation left, Ann chose to fly back to Madison with me.

From Billings to Sheridan, Wyoming the ground is at roughly 4,000feet above sea level. The temperature today was warmer by some twenty degrees than when I had flown this segment alone. Consequently, the density altitude was much higher. With two people and a lot more baggage in the Cub, 4,100feet was all the altitude it could maintain. If the horizon beyond a butte is visible, the Cub is going to clear the top of the butte. As we approached each butte, the horizon visible above the butte became smaller and smaller until it was obvious we were not going to be above it. Then began the struggle to gain enough altitude to achieve some semblance of clearance.

To get the Cub to climb a few feet, I would push the throttle to full power, pull back on the stick until the engine eyebrows were several inches above the horizon, and then watch the horizon to see if this had any effect on our altitude. Each time, the Cub would struggle up to about 4,100feet then the climb would just fade away. I would give full throttle again, weave side-to-side, gain a few feet, but the climb would fade away again. All-the-while, we would be edging closer and closer to the butte in front of us. The Cub cleared some of those buttes by no more than thirty feet. Making matters worse, the airport at Sheridan is some 400 feet higher in altitude than Billing's. This necessitated some additional climbing along the course line.

After an hour and a half's struggle to maintain altitude, we finally landed at Sheridan in the early evening. This time, the 93 miles took an hour and 45 minutes for a ground speed of 48 mph, which gave us lots of time to look at the buttes. A car was rented and Ann and I went to the local KOA campground. The woman running the place thought, it was quite silly to spend money renting a local car to drive to a campground. We could have stayed in a motel for the same amount of money. I thought it was perfectly rational, although Ann had her doubts. The camping gear had been lugged nearly 1,600 miles in the Cub. I was determined that it would get used as much as possible. We did use the car to visit the sights in Sheridan such as the Hotel Sheridan where Buffalo Bill Cody had stayed many times.

The next day (Thursday) dawned clear with light winds. Our first gas stop was Gillette. This time, I called the tower before we landed. The operator remembered the Cub as he made a comment about our using the radio. On the ground in Gillette, the realization of how close a call we had at the Lee Ranch came to Ann. She proceeded to write her thoughts in a diary. To this day, I have never seen what she wrote while sitting on the airport tarmac in the brilliant sunshine of a warm August day.

After checking the maps, my navigator decided that some time could be saved by not landing at Spearfish and going directly to Sturgis, South Dakota--106 miles away. This also bypassed Rapid City's control tower.

After taking off at Gillette, we deviated off the course line in order to fly close to Devil's Tower that was made famous by the movie Close Encounters of the Third Kind. The tower is a wide round column of grey rock 867 feet tall with numerous vertical flutes running from top to bottom. It sits out in the middle of open country on top of a low tree covered hill. In the movie, the alien's spaceship landed on top of the tower. From our vantage point, the flat top of the tower looked like an ideal place for such a landing.

Just north of the Sturgis airport is a small mountain, called Bear Butte, sitting in the middle of rolling ranch land. I do not understand why it is called a butte when its top is a peak rather than flat like a butte. Although the dictionary calls a butte “a steep hill standing in a plain.” None-the-less, it made a wonderful landmark as it could be seen from great distances across the prairie. I'm sure the wagon trains of the 1800's used it to navigate. The weather seemed to be cooperating as our ground speed on this segment was a satisfactory 62 mph. As the Sturgis airport's mechanic gassed up the Cub, he told me not to use cowl clips behind the propellor as they will gouge the back of the wood prop.

After a gas stop at Philip, Ann and I arrived in Chamberlain by late afternoon. The fellow attending the airport was a retired, 67 year old farmer whose tanned and weather beaten face reflected years of living out on the prairie. He had purchased a Cub to learn to fly and was quite interested in our trip. He revealed that he had taken some forty hours of dual instruction, but had not soloed yet. Even though one usually solos in about 10 hours, he wasn't the least bit discouraged and was looking forward to flying on his own. He gave us a ride in his car to the KOA campground on the south side of Interstate 90.

The next morning (Friday), having camped at the KOA proved to be a problem. It was too far to walk back to the airport and hitchhiking proved fruitless. Dozens of cars with high school students on their way to Chamberlain went by, but none stopped to pick us up. I can hardly blame them as both Ann and I had backpacks on and looked like we had been on the road a long time. To get back to the airport, we finally resorted to calling the town's one and only taxi. It was a dilapidated ten year old Buick with at least an inch of dust on the floor.

Now that we were back in the low country, It was Ann's turn to be the pilot. I could relax in the luxury(?) of the rear seat and serve as navigator. At Chamberlain we were back on our original flight path. The next fuel stop was at Mitchell only 63 miles away. Getting there took exactly one hour of flying time.

After Mitchell, the Worthington, Minnesota airport was out of the Cub's range, so we stopped for gas at the small town of Dell Rapids, South Dakota that is 15 miles north of Sioux Falls, South Dakota. It was a peculiar airport with one east-west runway going uphill to the east. Both sides of the runway were lined with heavy earth moving equipment that made me uneasy when we landed. The lineboy said that in a few weeks they were going change the airport's status to private. En route from Dell Rapids to Worthington, the clear skies became overcast and haze had started to develop. We were back in the Midwest again.

Ann and I arrived in Worthington just after lunch on Friday, August 25'th. JoAnn and her sister Clarice were surprised to see Ann. They thought Ann was going to return to Midland from Billings. We spent the evening recounting the events of the past few days. The weather was deteriorating rapidly and Ann had to be at work on Monday, so her “Cubbing” the rest of the way seemed unlikely. The Mullan Pass weather system was continuing to dog us. Reluctantly, I agreed that Ann and her mother would drive to Madison. From there she would take the airlines back to Midland.

They departed the next morning (Saturday). I strapped on the Cub to fly to Mankato and visit my parents. A highway and a railroad track go from Worthington to Mankato. There are seven small towns strung along them. Each of them a convenient horseback ride from the next. I had some of the worst weather of the whole trip on this leg. The ceiling was only 1,500 feet with lower scud clouds. There were large patches of ground fog scattered about. Although the country was flat, it was covered in corn fields. There weren't a lot of places to put the Cub down if the weather suddenly closed in. By diverting around areas where the visibility was too low and following the highway, the Cub and I were able to reach Mankato.

After an overnight stay with the folks, I departed for home with a gas stop planned at La Crosse. About a third of the way there, the weather became really rotten: low ceilings, less than a mile visibility, fog, and rain. I did the usual 180 degree turn and returned to the nearest airport, Owatonna, Minnesota. After sitting on the ground for an hour, I decided a nice bed at the folk's was preferable to a motel and flew the Cub back to Mankato. It was two more days before the weather improved enough to fly cross-country again. The route I took home went through Albert Lea, Minnesota and Decorah, Iowa. I arrived back in Waunakee on August 28, 1978. The Cub was put in its hangar and given a pat on the tail.

Even though the whole trip took only 11 days, it seemed like a much longer period of time. The Cub had been in the air 52 hours and had traveled 2,714 miles. Although it did not seem to be a problem most of the time, the overall ground speed was only 52 mph. That is slow when compared to other trips. The Cub had been flown over all kinds of terrain from the coulee's of Wisconsin to the mountains of Montana and the Great Plains in between. Ann was a splendid navigator. She tracked our flight on the map very closely, computed ground speeds every ten minutes or so and gave me ETA's to the next checkpoints. If we were even a few hundred yards off course, she would tell me how to get back on course. Providing the front seat person with maps contributed considerably to our ability to stay on course. Mechanically, the Cub never missed a beat or even shed a screw. Although our goal of Seattle was not achieved, a first class adventure was had by all.

Flight Record

SEGMENT	MH	Distance	TAC off	TAC down	
Wzunzkeo - Decorha Ia	270°	115 mi	983.74	985.47	[illegible] 1 43 4.2
Decorha - Forest City Ia	265°	95	985.47	986.88	5.7 1 24 4.1
Forest City - Worthington Minn	280°	101	986.83	988.29	5.9 1 24 4.1
Worthington Mitchell SDak	266°	123	988.32	990.07	6.3 1 43 3.7
Mitchell - Chamberlin	262°	64	990.07	991.20	5.0 1 08 4.4 1/2 way
Chamberlin - Philip	267°	113		993.2	8.3 2 00 4.2
Philip - ~~Spearfish~~ Rapid City	273°	113	993.23	994.1	4.8 57 (6.3) 1 35 [illegible] 4.6
Spearfish - Gillette Wy	250°	85	[illegible].51	997.08	6.4 1 14 (4.3) TWR 118.[illegible]
Gillette Sheridan Wy	278°	77	[illegible].03	998.51	1 26
Sheridan - Fort Smith Mt	295°	60	———	———	1 53 No Gas
Fort Smith - Billings	296°	44	998.51	1000.39	ATIS 126.3 TWR 124.2
Billings Livingston	254°	93	2 04		File Flight Plans 123.6
Livingston - Bozeman	FR	37	1002.43	1003.44 Filed	(FSS) 123.6 GAS 8.1
Bozeman - ~~Butte~~ Deer Lodge	262°	65	1003.44	1004.99	(4.5) 1 33
~~Butte~~ Deer Lodge - Missula	FR	116	1004.99	1006.10	CT 118.4 04 Deer Lodge Alternative
Missula - Kellog Idaho	FR	115			
Kellog - Henley	265° 336	23 20			
Henley - Odessa	227°	100	20 H		
Odessa Ellensburg	225	90	[illegible]		Over fly Moses Lake
Ellensburg - Bellevue	FR	85			
Missoula - Helena	FR	110	1007.77	Down	1 42 7.1 gal 4.2
Helena - Choteau	FPL	85	1009.20	← Down	1 26 6.1 gal 4.3
Choteau - Cut Bank	338°		047 1009.27	1010.15	53 4.1 gal
Cut Bank ∞ St. Marys	262/172			1011.77	1 31 / 1 hr 25 min 3.7 gal
Cut Bank - Great Falls	132°			1013.99	2 15 / 1 hr 25 min
Great Falls Lewiston	90°				36
Off. Fairfield		DNCO	1015.12		31
Lewistown - Billings			1017.23		2 20 1 06
Billings - Sheridan			1017.1[illegible]		[illegible] GAL 4.3
Sheridan - Gillette			1020.42		1 12

Church in Rural South Dakota

Wheat Fields in Western Montana

Glacier Park

Cub at Lee Ranch in Montana

Chapter 9
A Day Trip

In the summer of 1984, Ann quit her job of ten years with the Dow Chemical Company. She had found a deal where if you kept going east to some degree you could go around the world on one airline ticket. Ann was going to put all her necessities in a backpack and stay at youth hostels in the various countries. A week before she was scheduled to leave, Ann and I decided to take the Cub on a scenic trip to La Crosse, down the Mississippi River valley to Prairie Du Chien, and back up the Wisconsin River valley to Waunakee.

The fifth of July turned out to be a perfect summer day, blue skies, only a few scattered fair weather cumulus clouds, and little or no wind. As we climbed into the Cub, Ann said:

"Please Dad, no more adventures!"

As you know she had her share of them with me. She did not want a new one to interfere with starting on her trip. I solemnly told her that we were just going on a short day trip to enjoy the scenery and nothing untoward would befall us this time.

We took off around ten in the morning and started following the familiar route across the un-glaciated area to La Crosse. By now I knew every valley, creek and farm house along the route and really didn't need a map. Upon reaching the convent at St. Joseph's, we turned north to the Maple Grove Country Club, in the valley of the La Crosse River, where there is an east-west grass runway alongside the golf course . We landed and parked the Cub next to the clubhouse. After lunch in the restaurant, we spent some time watching the golfers. Since Maple Grove did not have fuel, we flew the few miles to the La Crosse airport and topped off the Cub.

After refueling, we took off to the south, climbed the Cub up to 1,000feet AGL, flew across the city, and then headed downstream over the middle of the Mississippi River. Just south of La Crosse, the river valley is at least three miles across. The river itself is about a mile wide and the rest is swamp. Tall bluffs line both sides of the valley as far as one can see. Once beyond the city, I thought about flying to the east side of the valley and following it to Prairie Du Chien. When we were five miles down river of La Crosse, a small thunderstorm could be seen some twenty miles further south. This time of year, such storms usually move along a southwest to northeast line. Because of this, I changed my mind, made a shallow turn to the west, and headed for the river's west bluffs. Hopefully, the thunderstorm would move northeast and out of our way.

A half mile from the bluffs, the Cub's engine started to shake violently and missfire. Looking down, I could see that we were directly over where the river melded into the swamps. A quarter mile to the west there was a tree-shrouded highway at the base of the tall bluffs, but no place to land.

I pulled back on the throttle and the engine shaking became much less violent, so I left it running. Over the years, I had decided that some engine power is better than none as it might be useful to stretch a glide or miss an obstruction. Since we had quite a bit of altitude, compared to our normal cruising levels, I decided to descend to the south along the west bluffs. Then, Ann caught sight of where the Root River joins the Mississippi River. An area that might offer some possibility for a safe landing.

As the Cub passed over the highway bridge that went over the Root River, Ann pointed out a 300-acre sod farm a half a mile to the west. She said we could easily glide to it. Once she knew we could make a good landing, she started to preserve the approach and landing on film. What a cool customer, but she has had quite a bit of experience in these sort of things. Landing on the sod farm was just like landing at Waunakee.

After landing and shutting down the engine, I pulled off the engine cowlings, but could see nothing wrong. As I was doing this, two college age fellows drove up on a large riding mower. We told them what had happened. They helped us push the Cub behind a building. Ann and I hopped on the mower and rode with them to the sod farm office. One of the fellows, called the owner, named "Barney," to tell him that he had an airplane down on the farm. The owner had a good sense of humor as his first question was "do they need sod?" He told the kids I could leave the plane where it was, so we all rode back to the plane.

As Ann and I were tying the Cub down, the local county sheriff's car came driving up to us in a hurry. Someone had called him reporting a crashed airplane on the sod farm. The workers said it was probably an old lady living up the road who was a busybody. I told the deputy sheriff that we had a minor engine problem and made a "precautionary landing." Whereupon he asked me for my license, so I gave him my automobile driver's license. He went back to the squad car and got patched through to the La Crosse control tower.

"There is an airplane down in a field."

"Is it damaged?"

"No."

"Where is it located?"

"Just west of the river."

"In what state?"

"Minnesota."

"It's not our problem."

While all this was going on, Ann and I ignored him and finished tying down the Cub. Having done his duty, the sheriff returned my driver's license, climbed in the squad car, and departed. I did not have the faintest idea of what he thought he was doing and I am not sure that he did either.

In the meantime, the owner of the sod farm showed up. He warned me that strong winds often came down the valley in storms and blew the irrigation pipe around. I decided that the Cub was reasonably well protected by the building and left it where it was. He gave us a ride back to the office, I called JoAnn and told her to come and get us again. Then, Ann and I walked to a tavern, a half mile away, that we had seen on final approach (even in an emergency, we pay attention to the important things).

Ann and I sat at the bar, ordered a couple of beers and were relaxing when a camera crew from the La Crosse TV station came in "has anyone seen the survivors of the airplane crash?" A couple of people at the closer end of the bar said that they had not heard anything about it. Ann and I solemnly shook our heads in agreement and could barely keep our composure. Fortunately, the TV crew left before we got the giggles.

After an hour or so, Ann and I walked back to the sod farm office. The college kids said that the TV crew had driven up all excited and were very disappointed to find a perfectly good yellow airplane parked neatly on the grass. The kids had the good sense not to tell them where we had gone.

In another hour, JoAnn showed up. On the way back home, we drove through the thunderstorm that providentially had caused us to move to the west side of the river. Had we continued flying to the east, the engine problem would have happened over the swamps. There was no sod farm on the east side of the Mississippi River.

Now, I had an airplane down in Minnesota with a sick engine and not a clue as to what was wrong with it. In any case, I could not do anything about it until the weekend. After thinking about the situation for a while, I decided the best course of action was to haul the Cub home. I started calling various friends to get some help with the hauling job. My Dentist and CFI, Dick Buescher volunteered to help. He insisted upon using his Ford station wagon that had a heavy duty trailer package for hauling his wife's horse trailer. It's a lucky thing he did as a few months later the frame on my 1969 Chevy station wagon rusted through and I had to junk it. I came very close to having both an airplane and car disabled in Minnesota.

Before we left, I bought a number of wood two-by-four's to build wing racks for the borrowed flat bed trailer. Both Dick and I brought a lot of tools along and several hundred feet of rope. Much to my surprise, Ann insisted upon coming along and helping, even though she was leaving on her trip in two days.

Upon arriving at the sod farm, the three of us started to take off the wings. The college kids working at the sod farm came along at just the right moment to help us. The only glitch occurred when the left wing was disconnected. I discovered the wires to the wing tip light were still hooked up. While one person kept the right wing from dropping, three other persons held the left wing in the air. I unwrapped the tape around the connection, cut it at the solder joint, and vowed to put in a good connector next time. Dick paid the College kids $10 for helping us. He would not let me repay him the ten dollars.

Once the wings were off, the three of us rolled the fuselage up onto the trailer using a couple of planks as ramps. The fuselage was securely tied to the trailer. Then we proceeded to build the wing racks. Dick had a battery powered drill and a bag of drywall screws that made the assembly job very easy. Ann took her turn at sawing the boards to the right lengths. After a couple of hours, the racks were done, we put the wings on the trailer, and tied them to the racks with rope. After tidying up the grounds, the three of us headed to Madison. Upon arriving at the Waunakee airport, the trailer was parked in front of the hangar and then we all went home.

The next morning, Burr Fontaine volunteered to come along and help me unload the plane. He and I took the wings off the rack, dismounted the racks, and rolled the plane off the trailer. By the time Dick arrived, the fuselage was in the hangar. While Dick and Burr held a wing in position, I installed the wing struts and bolts. This is straightforward, but you have to be very careful not to stress the strut forks. There have been a couple of in-flight failures of strut forks due to people cracking them when installing the wing struts. After reconnecting the wing tip lights, the wing fairings were put on and the Cub looked ready to go, except for one slight problem--the engine. The day after we brought the Cub home, Ann left on her trip around the world. JoAnn and I were not to see her for a year.

Bill Vogelsang, of whom we will hear more of in later chapters, and I started to look for the engine problem. The first thing Bill said was "pull the rocker covers." I poked around for a while and finally pulled the rocker cover off the left rear cylinder. I found that one of the keepers on a valve stem had worked up and the rocker arm had pounded it over the top of the valve stem--another accident waiting to happen. I decided to simply replace the keepers. I did and that cylinder has never given me another bit of trouble. I started the engine, but it still shook like crazy and made loud noises.

Clearly, this did not solve the fundamental problem. The next day Bill and I went up to look at the engine some more. Again, he said "pull the rocker covers." This time I heeded his advice and did so. When I removed the cover from the last cylinder, two chunks of the exhaust valve rocker arm fell to the ground. The arm had broken in two right at the center of the bushing. When the engine fired, the exhaust valve would not open, the burning gasses were trapped until the intake valve opened. Then, the exhaust gases went out the carburetor venturi. No wonder, the engine was shaking so violently. Since the underlying cause of the fracture was not apparent, I decided to take the engine home.

Once home, I started to disassemble the engine. After pulling off the offending cylinder, I could see the cup inside the hydraulic valve lifter had turned sideways and jammed. This effectively made the push rod about 3/16" longer and put an unusual force on the rocker arm eventually causing it to break. Strangely, the push rod was not bent.

In order to replace the lifter, the crankcase would have to be taken apart. When the engine had been overhauled a few years earlier, the mechanic had given me all the parts he had removed and I had saved them. Fortunately, several of my old lifters had perfectly smooth wearing surfaces and looked usable. I selected the best looking one, installed it, put in a replacement cup, and the original push rod. After that, it was a simple matter to have the engine reassembled. The next weekend, the engine was taken back to the hangar. Bill and I hung the engine on the Cub. After everything was hooked back up, I propped it and it ran perfectly. The replacement valve lifter has never given me a bit of trouble.

Looking back at the incident, Ann and I were lucky the thunderstorm caused us to veer toward the west bank where the sod farm could serve as an airport. Otherwise, we certainly would have landed in the river bottom swamps with unpredictable results. After the forced landing, it was essential to have friends to lend a hand when it was really needed. The humorous side is that now every time Ann and I climb into the Cub, her first comment is "Please Dad, no more adventures."

Chapter 10
Exploring Nebraska's Sand Hills

Shortly after moving to Madison in 1961, I met Bill Vogelsang (a faculty member in the Department of Nuclear Engineering) through the model airplane club. In the mid 1970's he bought an Aeronca Chief airplane. It is about the same size as a Cub, similar in construction, and is powered by the same 65hp engine. The major difference is that its two occupants sit side by side rather than in tandem. He restored it himself, earned his private pilots license at Morey's, and bought a hangar at the Waunakee airport. Being based at the same airport enabled Bill and me to go "airport hopping" together and attend many Fly-in's throughout Wisconsin. Over the years, Bill and I became quite adept at removing and replacing the Cub's engine.

Although Bill had grown up in Lincoln, Nebraska, he had never been to the Sand Hill country of central Nebraska. In the spring of 1985, Bill asked me if I would be interested in exploring it from the air. Ever willing to take the Cub on a trip, I thought it was a great idea. After some study of the maps, Bill and I planned a course for our two airplanes that would take us across northern Iowa to Yankton, South Dakota and then into central Nebraska via O'Neil, Nebraska. Once there, we planned to make a big loop through the Sand Hills before heading back to Madison via central Iowa. It would be a 1,298 mile trip. Since Bill was not into camping, we would stay in a motel each night.

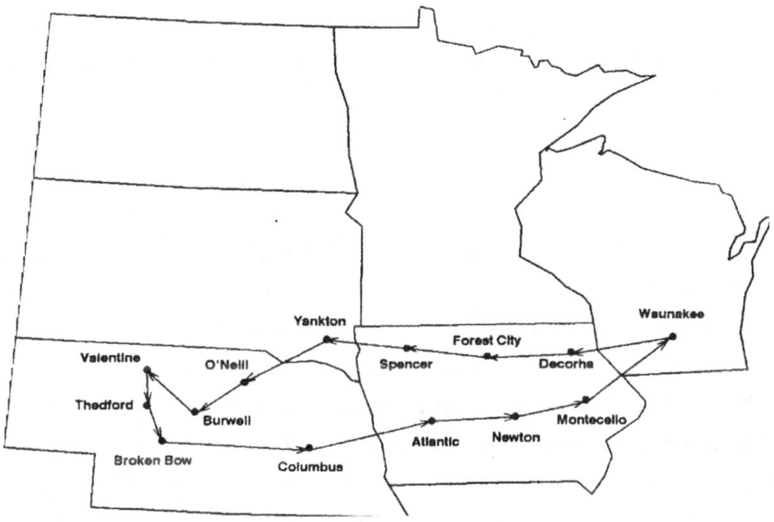

The Route of Flight: Sand Hills Trip

On Thursday, June 6,1985, the Cub and the Chief departed Waunakee at 8:00 A.M. and headed due west. Since I was the experienced cross-country pilot, I would act as leader and Bill would follow me. Despite Bill's Aeronca Chief being quite a bit wider than the Cub, it is about 5 miles an hour faster than the Cub. It soon became apparent that Bill would have trouble staying behind me. He quickly worked out a pattern where he would fly wide S turns behind the Cub.

For the first part of the trip, the sky was overcast and it was on the cool side. Northeastern Iowa is gently rolling with numerous valleys that transect the countryside on their way to the Mississippi River. Further west, the land becomes reasonably flat with green fields of corn or soybeans. In Iowa, all the highways are oriented either north-south or east-west. The only exceptions are roads that follow rivers. In the eastern half of the state the rivers run southeast to the Mississippi River. In the western half they run southwest to the Missouri River. Thus, it is difficult to get off-course in Iowa.

When we reached the northwestern part of Iowa, the sun was out, but the wind was 25mph and gusty. The direction and velocity of the wind can be determined quite accurately by looking at the fields below. The leaves of the corn and soybean plants are moved back and forth by the wind and they point in the direction the wind is going. The amount of waving of the leaves is a good indicator of the wind speed. The wind gusts superimpose a hundred yard wide pattern on the leaves that moves across the fields much faster than the basic wind. Sailors call these "Cat's Paws" as they can be seen as darker water creeping across the sea. By watching them, Sailors can adjust their sails to cope with the stronger wind in the gust before it arrives. In the Cub, the "Cat's Paws" on the fields are a signal that turbulence is to be encountered.

Having been alerted to the gusty conditions by the "Cat's Paws," both Bill and I made wheel landings at Spencer, Iowa. In that we had no particular destination scheduled, we waited for the wind to subside. A few hours before sunset, the wind dropped off and there was enough daylight left to fly to Yankton, South Dakota.

After remaining over night at Yankton's Holiday Inn, it was off to O'Neill, Nebraska. The airport there was one of those on a 1920's air mail route. It happens to be named John Baker Field. At a ground speed of 55mph, it took one hour and twenty minutes to fly the 73 miles. At O'Neil, auto fuel was available. The woman working at the desk said her insurance company would not let her pump it into an airplane, so Bill and I proceeded to top off our airplane's tanks ourselves.

I took off first and in a few minutes, Bill cut in front of me waggling his wings furiously. I tried to call him on my radio, but he could not understand a word I said and I was not receiving him. After a few minutes of this, it finally dawned on me that I was not on course. Bill herded me over to the correct heading for Burwell.

The countryside was beginning to look like cattle grazing land: lots of open country, clumps of trees here and there, and creeks wandering where we knew not. Navigating now became a matter of holding a compass heading until something shown on the map appeared. In the midst of this country, I flew low over a small pot hole filled with purple water. The edges of the hole were white sand with a strip of medium green grass next to the sand. Beyond this grass, there was dull beige colored range grass. About 100 Black Angus cattle were at the pot hole. Some were standing in the water while others stood on the banks.

After an hour of flying, the town of Burwell and its small airport, Cram Field, was sighted. Like most towns in this part of the world it was nestled in a river valley below higher hills. Near town, the Calamus River merged into the North Loup River.

After landing we found that the airport was unattended, but there was a telephone number to call for fuel. I called and in twenty minutes an old fellow appeared to refuel both planes. Cram Field is well known among antique airplane afficionados as it is where Dr. Cram keeps his collection of old biplanes. Bill and I were disappointed to learn the Doctor was out of town and that the old man did not have keys to the hangars.

Taking off from Burwell, Bill and I turned to the northwest and headed toward Valentine. This leg was 110 miles long through the east end of the desolate Sand Hill country. There were only a few north-south highways and several creek beds to help us stay on course. One of the few signs of civilization were the occasional metal windmills each with a water tank at its base. Most of these windmills also had a line of wooden power poles bringing electricity to the windmill's pumps. Due to the rolling nature of the landscape, these power lines just faded into the distance. Their source never was revealed. Cattle were widely scattered across the landscape. Most were brown and white Hereford, but there also were a few Black Angus.

Bill and I landed at Ainsworth, just to look around. To train B-17 and B-24 crews during WW II, the Army built numerous large air bases in Nebraska. The airport at Ainsworth was a classic example. The three runways were in the form a triangle and positioned in such a way as to face the prevailing winds. All but one runway had been abandoned. None of the WW II buildings remained. They had been replaced by a new metal airport office building staffed by an elderly woman. Bill and I talked with her for a while. She was well informed about the history of the airport.

After taking off, we followed a highway, a railroad, and a power line that all went from Ainsworth to Valentine. We landed at Valentine in mid-afternoon, refueled the planes, and tied them down for the night. The airport was close to town, so Bill and I walked to a nearby motel. As we checked in, the elderly lady at the front desk learned we were flying. She told us that the motel was on land that once had been an airport run by her late husband.

Bill and I had somewhat divergent views of how to dine when cross-country flying. I would want to eat in the local diner as inexpensively as I could. On the other hand, Bill liked to eat in the better restaurants. Fortunately, it became Bill's role to chose the restaurants. In Valentine, he found a four star restaurant named St. Christopher's. It served the most wonderful baking powder biscuits I have ever tasted. I ate so many of them I hardly had room for a meal. After dinner we wandered around town trying to visualize what it must have looked like when Bill's aunt taught school here during the 1930's.

In the rural towns of Great Plains, most houses are very small, need paint, and have front yards consisting of bare dirt with little wisps of grass. Many of these towns appear to be slowly dying. I was quite surprised at the nice neighborhoods, beautiful green grass lawns, and the general neatness of Valentine. Bill knew the town was the center of a major recreational area based upon the nearby Niobrara River. There were several companies in town offering canoe trips on the river. In the fall, Valentine is a hub for antelope hunting.

After breakfast Saturday morning, Bill and I walked back to the airport, loaded up the planes, and took off with Thedford as our destination. Two miles southeast of Valentine and across the Niobara River there is a major highway (US 83) that Bill and I planned to follow south to Thedford. When the Cub broke ground, I saw a highway ahead of me and started following it. After a few minutes I had not flown over the river and things did not seem right. Undaunted, the great navigator forged on. In another 3-4 minutes the Cub crossed a river and obviously was on course. Never mind that I was flying a heading of 240 degrees rather than 180.

About this time, Bill realized I was nowhere to be seen on the correct line of flight. He started looking around and saw me flying off to the southwest. Using his speed advantage he was able to get ahead of me and go through the wing waggling routine again. By the time Bill woke me up to my situation, I was nearly eight miles off course. A major change in the compass heading was made. Ten minutes later, both of us were above the right highway.

Eighteen miles south of town, we entered the Valentine National Wildlife Refuge. It was spring time in the Sand Hills, the range grasses were a medium shade of green, and there were occasional patches of white or yellow flowers. The land is gently rolling with depressions at the bottoms of the hills filled with either white sand or a pond of water. These ponds ranged in size from a few yards across to some medium sized lakes. I was not prepared for the amount of water that I saw. I expected the land to be nothing but prairie.

By this time a strong wind accompanied by considerable turbulence had developed. The Cub was getting tossed around and flying was becoming uncomfortable. To reduce the amount of the turbulence, I lost altitude until the Cub was only 10-15 feet above the highway. Bill also dropped down, but stayed off to my right. The north-south road went up and down over a series of hills that were oriented in an east-west direction. As I climbed over the top of one of these hills, I dropped the Cub's nose to go down the backside. Much to my surprise, I met a big semi-trailer truck, grinding its way up the hill. After looking the driver right in the eye for a split second, I pulled up sharply to get better clearance over the top of the trailer. I don't know who was more surprised, myself or the truck driver. I wonder what he would tell his fellow drivers about seeing an airplane come over the hill right in front of him.

By the time the Cub and Chief reached Thedford, the wind was blowing at 20 mph and was 45-degrees to the airport's single east-west runway. After my earlier experience at Escanaba, Michigan, I knew that I should land on the asphalt runway rather than on the dirt alongside it. Escanaba is where one of the Cub's wheels dropped into a deep hole hidden in the grass alongside the runway. After this earlier incident, I never taxi off the hardtop and am leery about landing on a grass runway that is not the main runway. I am a good one trial learner.

At Thedford, it took a well executed one wheel landing to get the Cub securely on the runway. But, taxiing in the strong wind was not a problem. I arrived at the gas pit just in time to watch Bill land on the dirt next to the runway. After he climbed out of the Chief, I mentioned that landing on the dirt was not a good idea. He felt that landing on the dirt would give him better control in the strong crosswind than would the asphalt runway.

Again, the airport was unattended, but there was a telephone number to call for fuel. In a half hour, a young fellow from the gas station, on the edge of town, arrived and refueled both airplanes. He said there was a good restaurant in town, however, it was not lunch time so we passed up the opportunity.

The next leg was from Thedford to Broken Bow, Nebraska. The latter town is located out of the Sand Hill country. There was a highway with a railroad alongside it connecting the two towns. Our course line was a few miles to the southwest of them. I got lazy and swung over to follow the iron compass to the southeast.

By now there was a high thick overcast and the world around us looked dark and gloomy. Despite this, the visibility was at least 10 miles. Just before lunch time, Bill and I landed at Broken Bow. Since the wind was increasing in velocity, both planes were tied down on the ramp. Then, we went into the airport lounge to wait and see what was going to happen with the wind.

One of the fellows in the pilot's lounge was the local undertaker who had been a P-38 pilot during the WW II. He told us some interesting stories about flying his Cessna 195 to isolated ranches to pick up the deceased. He took Bill and me to lunch at a local diner. It was interesting to hear about the local aviation scene. The airport was being run for the city by a retired air force sergeant who was a real rules and regulation guy. He had browbeaten the locals into believing that only the FAA anointed could touch an airplane. The undertaker was quite envious that both Bill and I had restored our airplanes ourselves.

By the time we were back at the airport, the wind had reached 55mph. Bill and I were thankful that the planes were securely tied down on the ramp. Roughly every half hour, I would call the Columbus, Nebraska FSS to check the weather. Eventually, it became clear that a dry front was passing through. It could be monitored by the wind velocities reported on each side of the front. By late afternoon, the wind at Columbus had dropped below 20mph. A gentle breeze was blowing here at Broken Bow. Bill and I decided flying conditions were acceptable. The planes were untied, engines started, and we took-off one after the other.

Broken Bow to Columbus was a 115 mile leg on a due east heading. In contrast to the Sand Hills country, here there were roads, railroads, and rivers cutting diagonally across the flight path. All of these ran from northwest to southeast toward the Loup River making for easy navigation. Along the route, we flew past a small river wending its way through the middle of a valley. The river bluffs were lightly wooded with trees having dark green leaves. Above the bluffs, the countryside was flat ranch land with the usual greyish green grass. A railroad trestle went across the valley and over the river. The trestle was similar in structure to the one that Ann and I had flown past in Montana. A quarter mile further on, a highway crossed the valley on a low concrete bridge. At one point, I passed over a rectangular shaped woods in the midst of the prairie that concealed a small town. There was a grain elevator along side the railroad track that went through town. I could see three or four stores along the main street. Most of the houses were hidden by the tall trees. By the time we arrived at Columbus, the sun was out, the clouds were gone, and the wind had died down. We landed and taxied up to the fuel pumps that I had used on previous trips.

The fellow running the fixed base operation is named Harold. Both he and his older helper owned J-3's. Harold gave us a long story about how he was leaving in a few weeks to attend the air show in Paris, France. He seemed impressed when I told him that JoAnn and I had been at Le Bourget airport only two weeks earlier on our tour of European air museums. He opened one of his hangars and both of our airplanes were rolled inside for the night. Harold gave me the keys to the courtesy car, so Bill and I could go to the Holiday Inn.

One of the problems of cross-country flying is that you get very tired from flying all day. For an airplane of its size the Cub is very light and consequently is susceptible to every quirk of the air mass. The plane moves in concert with every up or downdraft and wind gust. The latter will cause a wing to lift and the nose of the Cub to swing. The wind bouncing off the hillsides will create turbulence that causes the Cub to buck and weave. The throttle, control stick and the rudder pedals are used to counteract these motions. Thus, the Cub must be "hand flown" all the time to keep it straight and level, on the proper compass heading, and at the desired altitude. My arms and legs are in nearly constant motion. Most of the time, the motions are simply subtle pressures on the controls. Other times, the motions are major movements of the controls. In the midst of all this activity, I have to check the maps, keep track of where we are, and compute ETA's to the next checkpoints. Landing at the several airports along the route provides a small measure of relief, but until I reach my final destination for the day, the Cub must be strapped on again and be flown some more.

Another issue with the Cub is its uncomfortable seats that in a short time cause what is known as the "busted butt" syndrome. I resolved this issue by installing "Temperfoam" cushions in the Cub. They are made of a foam material designed for the astronauts seats in the Space Shuttle. As I sit on the cushion, my body heat makes the cushion conform to my posterior. This provides a great deal of comfort, but eventually even with these cushions my bottom goes dead.

The net result of all these factors is that, after a full day of flying, I am completely worn out even though I have been sitting the whole time. I really need to get some exercise. It takes some effort to take a walk at the end of the day. Fortunately, Bill insisted that we do so each evening.

The next morning, Sunday June 9, 1985, the weather was clear and the dry front had moved far enough to the east that it was no longer a problem. We pulled the planes out of Harold's hangar, loaded up, started engines, and took off for Atlantic, Iowa--118 miles away. The route of flight took us over the city of Fremont, Nebraska, that is just north of Omaha. As the Cub crossed Interstate 25, I recognized a red roofed Nickerson Farms restaurant where JoAnn and I once had lunch. In Iowa, just across the Missouri River, there were several very tall television towers right on the course line. Visibility was good, so Bill and I gave the towers a wide berth.

I wasn't quite so careful a while later. I was flying just above tree top level. As I came to a small town in Iowa, Bill called me on the radio to inform me that I was about a half mile north of course. As I initiated a turn to the right, I saw a radio tower that was exactly on my former heading. Its top was just above the trees and at my eye level. If I had continued on my former heading for a few more seconds, I would have had a close call with the tower--shades of Dodge Center.

With regard to radio towers, the best plans are to fly low enough that they stand out against the horizon or high enough to be above the tallest one on or near the course line. I prefer the former plan, the worst case is to fly so that the top of the tower is even with the horizon, which is what happened here.

The leg to Atlantic took two hours and five minutes of flying time. After we refueled the planes, I walked past the front of the Cub. I noticed the vertical tube on the carburetor air box was at an angle. Upon closer inspection, I found that the weld at the top of the tube had fractured and only a short section metal was holding the tube on. I pulled the tube off and stored it in the baggage compartment, so it could be reattached when we got home.

I had been at the Atlantic airport on a previous Denver trip. The runway arcs up to the northwest and there is a low railroad embankment just beyond that end of the runway. There also is a small hill a short distance beyond the railroad tracks. The combination of these three things makes it appear as if there is a high hill to climb over on takeoff. Today the wind was from the east, so the strange visual effect did not enter into the takeoff. From Atlantic to Newton, Iowa it was an hour and a half flight on an easterly heading. Bill and I landed at Newton (the home of the Maytag Company), refueled the planes, and took off for Monticello, Iowa.

The Monticello airport has a long northwest-southeast asphalt runway and a short east-west grass runway. After landing at Monticello, We had to taxi up a rise to where the hangars, the airport office, and the gas pit were located. It was a bit strange to look down at the runway from the gas pit.

After refueling the Cub and the Chief, Bill and I headed out on the last leg of the long journey. As soon as we were airborne, Bill abandoned his tail-end Charlie position and passed me. The Chief looked like a horse headed for the barn. In a few minutes it disappeared to the northeast.

The route of flight to Waunakee took me along the northwest side of Dubuque, Iowa where there is lots to see. Just before the town of Peosta, a four-lane highway (US 20) is crossed. It is the same highway that goes past the Chadron, Nebraska airport. The buildings at the Dubuque airport can be seen at about six miles to the southeast and across another wide valley. At Dubuque, the four-lane highway goes downhill through a steep cut in the Mississippi River bluffs. The highway crosses the river on a tall multi-span truss bridge and then climbs the bluffs on the east side of the river. On the west river flats, south of the bridge, there are many large white petroleum storage tanks.

Once across the river, it was still 75 miles to Waunakee. This was country that I had flown over many times and there is a string of small towns along the highway to Madison. In a few minutes, I could see the tall Blue Mounds hill that is a good aiming point as it is right on the course line. Once past, Blue Mounds, I was in my local flying area and put my maps away. By the time the Cub arrived at Waunakee, the Chief was in its hangar and Bill's car was gone.

The 1,298 mile trip took 21 hours and 20 minutes of flying time at an average ground speed of 61 mph. Although Bill and I had flown over some country I had seen on other trips, it was still enjoyable. Having a second pilot helps keep the great navigator on course. Each night, I was able to get Bill's perspective of the day's events. We both concluded that the trip was a lot of fun and we would do another one.

Chapter 11
Across the Great Smokey Mountains

Three years after the trip to the Sand Hills, Bill and I were thinking about another Cub/Chief trip to a different part of the country. My long time friend Gary Witt had recently moved his auto parts manufacturing business from Los Angles, California to Gastonia, North Carolina. It was far enough away (739 miles) to be a long trip. In addition, we would have to cross the Great Smokey Mountains to get there. Bill and I decided that en route we could work in a visit to the U.S. Air Force Museum at Dayton, Ohio. Bill purchased eight maps (Chicago, Detroit, Kansas City, St. Louis, Cincinnati, Memphis, Atlanta, and Charlotte) for each of us. In the cockpit, a large number of maps is a nuisance as the line of flight often goes off one map and onto another. This necessitates folding the current map, putting it away, getting out another, unfolding it, finding your current position, and writing the time for the next check point on the map. All the while trying to keep the Cub on the straight and level.

Route of Flight of the Gastonia Trip

Bill and I departed on Monday, May 30, 1988 at 8:00 A.M.. After a gas stop at DeKalb, Illinois, the next stop was at Rensselaer, Indiana. As I paid my fuel bill there, the old fellow running the airport office asked me where we were from. I told him Wisconsin. He asked if I knew Bill Coleman. I inquired as to how he happened to know Bill. He said that it was because a Portage, Wisconsin area pilot had been awarded a Wright Brothers trophy for his homebuilt Pietenpol airplane. The pilot was flying it to Dayton to receive the award. Bill and his wife, Lou Ann, each had flown their own Cubs along with him. Near Rensselaer, the Pietenpol developed magneto problems, so they landed there for repairs. It took a day or so for the necessary parts to arrive. While waiting for them, Bill and Lou Ann spent considerable time at the airport.

The present spring was an unusual one in that a large stagnant weather system was parked over the Midwest. The temperatures had been in the mid-nineties for over a week, the humidity was very high, the winds were light, and there were no clouds. In the resulting heavy haze, the visibility was less than 5 miles. The high temperatures and high humidity were a perfect recipe for magneto troubles. Sure enough, the Chief's decided to act up. Bill and I worked for an hour trying to cool them down, but to no avail. Finally we rolled the Chief up next to the airport office and used a hose to run cold water on the engine. The water was the foulest smelling stuff I have ever encountered. I asked the airport manager how they drank it. He replied they don't--they drink bottled water. Despite the smell, the water cooled down the magnetos. After a delay of some three hours, the Chief's engine was running again. The Cub started easily, so Bill and I took off and headed southeast over the flat Indiana countryside that is mostly corn and soybean fields with occasional groves of trees.

The next stop was Alexandria, Indiana, a small airport where a warbird rebuilder is based. He had a nice T-28 parked on the ramp. In the hangar there was another one under restoration. After Alexandria, the destination for the day was Montgomery County Airport some 10 miles south of Dayton, Ohio. Bill and I landed there just before sunset, tied down the planes, reserved a room at a motel, and rented a car.

Tuesday was spent visiting the U.S. Air Force Museum. It has the world's largest collection of military airplanes and includes planes from WW I to the present. A huge new third building was behind the two original buildings. To me, the most interesting plane this building was a Douglas O-46A, a 1930's Army observation plane, that was suspended from the ceiling. It had a blue fuselage and yellow, elliptical, parasol wings. During the 1930's and 40's, rubber power free flight models of this airplane were popular.

The sections of fabric from WW I fighter planes were the most unique displays in the museum. At the end of the war a number of pilots simply cut the fabric, containing their squadron and personal insignias, off their aircraft. They took the pieces home as souvenirs. Somehow these fragments eventually made their way to the museum.

Bill and I went over to the Museum Annex on the far side of the field. There was not a lot of interest there, but on the ramp there was an absolutely perfect WW II German JU-88 bomber. It looked as if all one had to do was climb into the cockpit, hit the starters, and go flying. This plane was new when a Rumanian pilot defected with it near the end of the war.

Before taking off on Wednesday morning, I checked the weather with the local FSS and asked about the military operations area (MOA) on our route from Dayton to Ashland, Kentucky. The briefer said it was active with F-4 Phantoms flying through it at 100 feet above the ground. Not wanting to mess with an F-4, Bill and I planned to deviate around the MOA.

After taking off from Dayton, there was not a cloud in the sky, but the visibility was no more than two miles. The heavy haze made accurate navigation even more critical. Fortunately, there were many check points along the course line. At our low altitude, I could not see them through the haze until I was right on top of them.

Ashland was 110 miles from Dayton. About half way there, the countryside changed from flat farm land to multi-fingered low ridges and valleys. There were gravel roads working their way up and down the ridges and through the valleys. The Cub and Chief crossed the Ohio River a few miles southwest of Portsmouth, Ohio and then entered Kentucky. From there it was only 15 miles to Ashland. Its airport was on the south side of the Ohio

River between the water and raised multiple railroad tracks that were tight against high river bluffs. To land, we had to make our traffic pattern next to the south bluffs and over the railroad tracks. All of this gave me the uncomfortable feeling of being hemmed in. Several corporations base their jet aircraft at this airport. I'll bet that their pilots have second thoughts about making an instrument approach down into the Ohio River valley.

Not knowing what awaited us further along the route, Bill and I walked a mile or so to a small food stand and had lunch. The part of town we walked through looked to us like late 1930's. There were many small rundown houses with peeling paint. The street was in need of repair. The few stores clearly had seen better days. This part of town simply looked dilapidated and grimy. After lunch, we went out to start engines and the Chief's was balking again. We propped it for a while and it finally coughed to life. Now it was my turn, as the Cub decided to act up. After flipping the prop for about five minutes, I finally got the engine started.

After takeoff, Bill and I flew north in the river valley to gain some altitude. Once high enough, we took up a heading of 173 degrees for Lonesome Pine, Virginia--114 miles away. How could we not land at a town with such a name? When clear of the river, the Cub and Chief started flying over some of the worst country I have ever seen. It was a solid mass of tall, sharply peaked, wooded ridges with small creeks at the bottoms of narrow valleys. Usually, a dirt road wove in and out of the trees alongside of the creek. The sides of these ridges were at a 45 to 60 degree angle and were covered with hardwood trees from top to bottom. The narrow tops of many ridges had been stripped mined. They were a jumbled mess of holes, piles of dirt, and embankments. If engine trouble developed, there was absolutely no place to land safely. The only comparable country was the lava flows on the side of Navajo Mountain at Lake Powell in Arizona. I think one's chances of making a reasonable landing were much better on the lava flows.

As I flew over each of these ridges, only a grey wall of haze could be seen ahead. In a minute or so, the next ridge line would become faintly visible. I would have no idea if it was higher or lower than my altitude. To me, they all loomed out of the gloom at a much higher altitude than I was flying. As the Cub approached them, they would appear to sink and eventually slide silently beneath, only to reveal yet another valley with a creek at its bottom.

Near the end of this leg, the ground level started getting higher with the Lonesome Pine airport at 2,670feet. Approaching Lonesome Pine, I could see the airport buildings on the high ground, but could not see the runways. The whole area had been strip-mined and had the appearance of a moonscape. Somewhere in the piles of dirt was a runway. Strangely, on a wooded ridge across a narrow valley from the airport was a neighborhood with many expensive homes. Finally, the runway came into view. I could see that there was a drop-off into a deep valley at the approach end of the runway. I added some altitude to my final approach to compensate for the inevitable downdraft.

After landing and refueling the airplanes, Bill and I went into the airport office. We talked to a helicopter pilot, who was familiar with the mountains. He showed us what he thought were the best routes to Morgenton, North Carolina. All of the routes were too circuitous to be of much value to us. Our decision was to remain on our planned route straight to Morgenton a 105 miles away on a 145 degree heading.

The first 30 miles or so out of Lonesome Pine were over mountainous terrain. Then, we came to a wide valley that contained three small cities: Kingsport, Bristol, and Johnson City, Tennessee. As we entered this valley, I pulled one of my old tricks by flying toward a landmark on the horizon. After 10 minutes, I started feeling uncomfortable. I was finding plenty of check points, but they didn't fully agree with the map (Valentine again). There would be a road and a railroad track, but there should have been a bridge to complete the checkpoint. After about a half a dozen of these "almost" check points, I decided that I didn't really know where I was. I talked to Bill on the radio. He surmised we were left of course. I made a 45 degree turn to the right, and flew on the new heading for ten minutes.

Finally, we flew over a large airport, but there was not a single sign on the ground to identify it. In addition, there was not a person in sight. After conferring via radio, Bill and I decided the airport was at Johnson City. It was about 10 miles to the right of our course line. I took a new compass heading and after a while was able to get on our original course line. Even though we had found an airport, this was not the kind of country in which to get confused. A few miles off course in the two to three mile visibility with 10 or 15 minutes of no checkpoints, I would have been hopelessly confused (navigators are never lost, only momentarily confused).

As Bill and I flew south, the visibility gradually improved to about five or six miles. When the last ridge appeared, it was very high. In addition, a 5,940foot peak was right in our path. To its left was a very wide saddle in the ridge line that was much lower than the peak. As the Cub flew into the saddle, I could see numerous lodges, summer homes, and recreational facilities tucked in the woods. In the summertime, people who live in the Piedmont of North Carolina come here to escape the heat and humidity.

At the eastern edge of the Great Smokey Mountains there is a precipitous drop-off. Only 22 miles away, the Morgenton, North Carolinia airport was at 1,266feet, a difference in altitude of 4,674 feet. Once over the saddle, I began a gradual descent and in a few minutes was over farmland. I was relieved to finally be out of the mountains. By the time Bill and I had dissipated most of the excess altitude, we entered the Morgenton traffic pattern. I landed after a Cessna with Bill right behind me. The airport was a modern facility with many parked airplanes. There was considerable traffic for so late in the day. It was quite a contrast to many Midwestern airports that have little traffic and seem to be slowly dying.

It was only 50 miles from Morgenton to Gastonia, North Carolina. As Cub and Chief approached Gastonia there were two tall television towers to the left of course with their red lights blinking in the dusk. Just like the towers in Madison, they looked as if they would reach out and grab the airplanes. Bill and I gave them a wide berth as we looked for and finally located the Gastonia airport. After landing, the airplanes were tied down, and we unloaded our gear. The 739 miles required 11 hours and 50 minutes of flying time at an average ground speed of 62mph.

The fellow at the airport office gave us the keys to a beat up airport courtesy car. Bill and I proceeded to try to find Gary's house. Much to our surprise, he lived only a block or two from the airport. Not getting any answer at the front door, we walked around the house to find Gary and Bev lounging at their swimming pool. Bill and I were a day early on our ETA, so they were not expecting us.

Bill telephoned his technician at the University and instructed him to take the magnetos off Bill's spare engine, and ship them to Gastonia via Federal Express. The magnetos arrived the next day. Bill took them out to the airport to have the local A&P install them. Upon closer inspection it was determined that in order to mount the new magnetos, the mounting studs on the crankcase would have to be replaced by longer ones. The A&P did not recommend doing that task this far from home. Reluctantly, Bill put the new magnetos in the Chief's baggage compartment. The strange thing was that the original magnetos worked perfectly all the way home. They knew their days were numbered if they did not perform.

Thursday afternoon, I took Gary and then Bev for a ride in the Cub. Gary had been one or our high school "fighter pilots" and had not lost his touch. Bev had not flown before, so I gave her the usual "this is how the controls work" routine and in a few minutes she was making nicely coordinated turns. A few years later, they came to Madison and I went with them to the EAA convention at Oshkosh. Bev became very enthusiastic about aviation. Once home, she quickly earned her Private Pilot's license. Gary received his about a year later. They bought a Cessna 172 and have become experienced cross-country pilots. They have flown their plane to California, Wisconsin, Minnesota, Montana, and Nova Scotia, Canada.

On Friday, June 3, 1988, Bill and I packed up the airplanes, said our goodbyes to the Witts, took off, and headed southwest toward Anderson, South Carolina. The plan was to fly around the southern end of the Great Smokey Mountains and then go to the northwest. On the way to Anderson, I was surprised by the number of large cities along our line of flight. We had to fly south of several of them to avoid control tower zones. Although it had worked well earlier, my radio now only worked intermittently, so Bill did all the radio work. The airport at Anderson was large and had an FSS.

In the airport pilot's lounge, I talked to an old fellow who owned a Piper J-2 Cub. The previous summer he had flown it right up the spine of the Great Smokey Mountains to the Piper Fly-In at Lock Haven, Pennsylvania. Once there, he should have greased the exposed rocker arms on the Continental A-40 engine, but neglected to do so. On the way home, he was in the middle of the worst of the West Virginia ridge and valley country when the engine froze. He had some altitude, spotted a hayfield ahead, and set up an approach to it. As he glided lower and lower, the field came closer and closer. Finally, he recognized that it was on the side of a 45-degree slope. By this time, he was committed to landing. As the Cub touched down, with a heavy side-wise component, the right landing gear collapsed. The propeller broke when it hit the ground. Now, he had a real problem as the incapacitated Cub was resting on the side of a ridge several hundred miles from home. It would have to be trucked out. It took him a couple of weeks to get back to the site with a trailer, but he got the plane home. Inspection revealed that the landing gear was nearly rusted through and its breaking was only a matter of time.

Bill and I went into the Anderson FSS to check weather. Since this was one of the FAA's new "super" FSS's, I expected to see some high technology, but their video screens were the same ones I see on the TV's Weather Channel. After checking the weather, we climbed back into the planes and taxied out. Just as I lined up on the runway, the FSS briefer called Bill and told him there were severe thunderstorms moving east to west along our flight path. They would reach Anderson in about 15 minutes. Since we were to the west of the storms, Bill and I decided to takeoff and try to keep ahead of them. I hoped that this would not turn out to be another LaCrosse experience.

The skies were completely overcast with a dark grey ceiling at 1,500feet AGL. The visibility was only four miles in heavy haze. It was so mucky that I could not tell what I was flying into. Normally, when approaching thunderstorms, you can see a nearly black wall of clouds. Unfortunately, this line of storms was behind us. Looking back toward the east, I could not see them in the gloom. I had a queasy feeling that they were catching up to us and would strike with no warning.

Now, the Cub and Chief were flying over solid woods. Bill and I could go for 20-30 minutes at a time without seeing an open field. We flew over several very large industrial complexes sitting in the middle of the woods. I never did see where the workers lived, but it was not near the factories. When I was in the Air Force, some of our training missions would take us across the southeastern part of the U.S. Being from the Midwest, I expected to see nothing but fields of cotton. Much to my surprise, the countryside was mostly pine forests.

It took us two hours and 10 minutes to fly the 99 mile leg from Anderson to Jasper, Georgia where the sun was shining. Bill and I had a ground speed of only 46 mph on this leg that gave us plenty of time to fret about the storms behind us. The airport at Jasper was on the top of a ridge. As we taxied up to the gas pit, there seemed to be an unusually large number of people around the airport office. We learned that the local EAA chapter was having a Fly-in the next day. The people were preparing for the event.

Bill and I talked to two fellows, both named John Murphy. One was an older man who had owned a J-3 since 1954. He only flew it when there were clear blue skies and a dead calm. The younger John Murphy, sans a southern accent, appeared to be someone who had retired to Jasper. We thought he might have been ex-military. He told us the weather forecast was for severe thunderstorms later in the day, so Bill and I decided to stay over night.

Both planes were rolled behind one of the hangars. We attempted to screw the tie-down anchors into the ground, but they would not go in. The soil was typical southern red clay filled with pieces of red shale. The tie-down anchor would screw in about an inch, hit a piece of shale and stop. After a half hour of making shallow holes, Bill and I finally screwed the tie-downs for both planes far enough into the ground to feel comfortable about their ability to hold in a strong wind.

The younger John Murphy called the local motel, arranged for a room, and drove us into town in his pick-up truck. The town sat in the valley below the ridge. The motel was very nice with an excellent restaurant. Since my nerves were shot from being chased by unseen thunderstorms all day, I treated myself to a nice meal. Later, Bill and I wandered around town for a while. It was a typical small southern town with its name in black letters on a silver water tower. The town appeared to be on the edge of being prosperous, but didn't quite make it.

The next morning (Saturday), young John Murphy picked us up at 8:00 A.M. and took us back to the airport. Bill and I were relieved that no thunderstorms had gone through during the night. The day's Fly-in festivities were just beginning. A few of the early birds had already landed, so Bill and I thought it would be best if we departed as soon as possible.

The next flight segment was to Winchester, Tennessee --104 miles away on a heading of 298 degrees. En route, the countryside changed from pine forests to a mixture of woods and fields interspersed with gentle ridges. When we flew southwest of Chattanooga, I could see the city at the end of a valley. About this time, I started thinking my engine was running rough. The more the Cub climbed to clear a ridge, the rougher I thought the engine sounded. There was a small private airport off to the left. I thought seriously about landing there. Finally, I decided the problem was "automatic rough," which is purely psychological.

Twenty miles further along our path, we approached a high, tree covered, southwest-northeast ridge from its south side. The northwest side of the ridge had a steep 1,000 foot drop off. Several large mansions were located along road at the top of the ridge. Bill informed me that the ridge was Cemetery Ridge, the site of a major battle during the Civil war. I certainly would not want to have been the army down below the ridge.

By now, it was a beautiful sunny day. The Cub and the Chief were moving along the flight path nicely. When Winchester was reached, the Cub had enough fuel to fly to Tullahoma, Tennessee, only 17 miles further. Approaching Tullahoma, the large NASA Arnold research facility could be seen in the middle of some woods just like the factories in Georgia. The Tullahoma airport had been a military base during WW II and some remnants of its past could be seen. At one time, the local EAA group had tried to hold a major regional Fly-in here, but it failed to become established.

As Bill and I walked around the airport, we saw a number of homebuilt airplanes in the long rows of hangars. There also was a WW II Brewster Buccaneer airplane sitting alongside the road. Every so often, a picture of this plane appears in one of the aviation magazines. A few years later, I read that someone had removed it and was going to restore it to flying condition--a major task.

One of the reasons Bill and I wanted to stop at Tullahoma was to visit the Staggerwing Museum. We walked over to it, but it didn't open until 1:30P.M. Since it was only 10:30A.M., we did not want to wait for it to open. Bill and I had to be satisfied with peeking through the windows into the dark rooms. The Beech Staggerwing and Travelair biplanes that we could see appeared to be very interesting.

The next airport on our itinerary was at Dickson, Tennessee. We took off from Tullahoma and soon were flying over countryside that gradually changed into low rolling hills. This part of the world was settled by the English who did a good job of replicating their homeland. The farm fields were relatively small, irregular in shape, and each was surrounded by a hedge row. Some fields were a tan color, others were a subdued green that was duller than the lush green fields of Wisconsin.

By the time we landed at Dickson it was nearly noon. The lady in the airport office said there was a gas station/convenience store about a mile down the highway. Being hungry, Bill and I walked to it. We bought the most appealing of their junk food for lunch.

The next leg was to Sikeston, Missouri and the flight turned out to be quite interesting. Thirty miles northwest of Dickson, Bill and I flew over a man-made reservoir called Kentucky Lake. It was about two miles wide and stretched to the north and to the south as far as we could see. Some sixty miles further, we crossed the Mississippi River. The land here is very flat and the river is a wide brown ribbon cutting through it. Due to several years of drought, the water level was very low and there were numerous sandbars in the river. The barge captains would have to be very careful to avoid running aground.

Sikeston was 25 miles northwest of the Mississippi River in what millenniums ago had been the river bottoms. The airport was on the northeast edge of the city and had been a WW II air base. It had a single long hard-surfaced north-south runway. After landing and topping off the airplanes, Bill and I talked with some of the locals. We learned that the airport once had been very active with several rows of airplanes tied down alongside the runway. Now, the only traffic was due to several local corporate airplanes. They also mentioned that a group of EAA chapters had established their own grass strip airport northwest of town.

Leaving Sikeston, The Cub and Chief were put on a more northerly heading(320 degrees) towards Farmington, Missouri. Twenty-five miles out of Sikeston, we flew past the EAA chapter's airport. It had a single large hangar and one grass runway. Other than being in the middle of crop fields, it looked like a nice setup. Once past this airport, it was mostly forests to the left of course. To the right of course there were groves of trees, green farm fields, and quite a few farm houses. The weather was nice, so it was just a matter of grinding out the 76 miles in an hour and five minutes (70mph).

After refueling at Farmington, Bill and I headed for Rolla where Missouri A&M University is located. By now it was late in the afternoon. The Cub and Chief were flying on a westerly course (280 degrees). There were a pair of good checkpoints in the first 10 miles. From these, a good wind correction angle was established. After these checkpoints, Bill and I started to fly over wooded country interspersed with small creeks and an occasional dirt road. Initially, I didn't think much about it as I have flown over a lot of sparsely populated countryside.

The further we proceeded along our course line, the less civilized the land below became. I was not prepared for the kind of country we were flying over. I had expected a continuation of the typical farmland we had seen up to Farmington. The countryside beneath us was similar to the swamp lands between Iron Mountain, Michigan and Three Lakes, Wisconsin, but not as wet. The only checkpoints, in some 40 miles, were a power line and the hamlet of Cherryville. Incongruously, in the middle of this desolate country was a large lodge. It appeared to be a vacation destination. Hiking and hunting were about all one could do there.

On the map, Rolla was just at the edge of the paper. Without using two maps, I could not tell what was north of our course line. Finally, a small community could be seen to the right of our line of flight. I flew over there to read the name on the water tower. Once there, it was determined that we were a few miles off course, but could follow an Interstate highway directly to Rolla. The main airport for Rolla was 13 miles north of town. After some discussion via our radios, we thought that airport could present some problems in getting to and from town, so Bill and I opted for the Rolla Downtown airport.

This airport was another of those depression era looking airports. It was located on top of a hill overlooking the town. It had an east-west runway, a long row of dilapidated rusting metal T hangars, and the grounds were overgrown with weeds. We landed to the east, tied down the planes in the weeds, and walked to the office, but it was closed. We had reconciled ourselves to walking to town when a Cessna 172 landed. A fellow and his wife climbed out. Bill and I hung around our planes until they offered us a ride to the nearest motel.

Just a block away from the motel was a typical Midwestern steak house. The couple said that it had a reputation as being one of the best in Rolla. Not one to miss a good meal, Bill and I went there for dinner. After being seated, I looked around and noticed that every man in the restaurant was wearing a baseball cap. Some were obviously from a softball team and were eating after a game. Others were just locals out for dinner. Bill and I got a silent chuckle out of the attire. Ever since then, whenever I see men with their caps on in a restaurant, I am reminded of gracious living ala Rolla.

The next morning, Sunday June 5, 1988, the weather was beautiful, blue skies, light winds, and a comfortable temperature. It also was to be our fifth day of flying. The taxi ride to the airport cost $10 for a five minute ride. The cabby justified the cost by saying that the airport was out of town. I was glad we did not land at the main Rolla airport some 13 miles north of town. Since we had not been able to refuel the planes the night before, we needed gas. One of the local pilots was waiting at the gas pit for fuel. He said the fellow would be along shortly. Bill and I rolled the Cub and Chief in behind his plane and waited. By 9:00 A.M., the planes had full gas tanks, we were back in the air, and headed for our first gas stop at Sedalia, Missouri. To the south of us, I could see tree-covered hills---the Ozarks. We also flew about 10 miles from the Lake of the Ozarks, but it was hidden behind some hills.

At Sedalia the airport office building was on top of a small rise overlooking the ramp and runway. Bill and I quickly refueled our planes, took off, and headed for Kirksville, Missouri--99 miles to the north-northeast. En route, we were flying over rolling land with creeks in the valleys and some trees on the hillsides. This countryside was devoted primarily to grazing cattle rather than growing crops. The grass was a dusty green color and was uniformly close cropped. At one point, I saw some fenced in areas with animals that I could not immediately recognize. It took me a few minutes to figure out that the animals were deer and elk in a game farm. Compared to the Farmington-Rolla leg, this landscape was quite civilized.

En route, the Cub and Chief crossed the Missouri River that was working its way in lazy S turns to the southeast where it would eventually join the Mississippi River. The water level was only a few feet below the tops of the sharp edged low river banks that were capped with a narrow row of trees. In places, this gave the river the appearance of being a canal. The whole area adjacent to the river looked as if it would flood easily. The Missouri River was on a considerably smaller scale than the Mississippi, but still impressive.

The rest of the flight to Kirksville was uneventful. There were roads, small towns, rivers and power lines for check points. The nice weather and good visibility made navigating easy. All I had to do was wait for the checkpoints to slowly work their way from the horizon to underneath the Cub. Upon reaching one, I calculated the ETA of the next checkpoint and this process was repeated many times.

Because the terrain was rather sparsely populated, I reverted to flying the Cub only a hundred feet off the ground. There was little or no wind and the skies were clear, so tooling along just above the widely spaced trees was an enjoyable way to spend the time.

Kirksville has a number of tall white water towers that can be seen from quite a distance. They tried to lure me into flying toward them, but they are five or six miles north of the airport. Fortunately, the airport sits on slightly higher land and can be seen from afar. Thus, the siren call of the water towers was for naught.

The runway's 6,000 foot length meant either a long taxi after landing or flying along the runway at 10 feet until the Cub is only a few hundred feet from an off-ramp and then landing. Normally I chose the latter technique as in windy conditions it minimizes the taxi time. Bill and I parked our airplanes on the large ramp and waited for the lineboy to dispense 100LL fuel from a gas truck.

At one time, commuter airlines flew into Kirksville, so there was a small airline terminal building. Since it was 11:30 A.M., Bill and I stuffed the terminal's junk food machines with coins to get enough chips and candy bars to kill the hunger pains.

From Kirksville, the line of flight was northeast on almost a straight line back to Madison. The first stop was Washington, Iowa--93 miles away. Fifty-five miles from Kirksville the Cub and Chief crossed the Des Moines River that was much smaller than the Missouri River. There were two quaint villages on the river's edge. Leando on the south bank and Douds on the north bank. They were connected by a brand-new concrete bridge. Each village had a number of large Victorian houses and a white church with a tall spire. The valley was quite narrow, but the bluffs were only a hundred or so feet high. From there to Washington, it was corn fields all the way.

At the Washington airport, the office was open, but unattended. Bill and I had visions of being stuck here for the night. Finally, a young fellow wandered in, he told us they had an air show in the morning. The FBO had been providing local people with airplane rides all day and had gone home to rest. After some cajoling, the fellow figured out how to unlock the fuel pumps and we refueled the planes. He had no way to make change, so I solved the problem by putting the fuel bill on my credit card.

After Washington, the next stop was Savanna, Illinois. Its Stransky airport is located on the Mississippi River bottoms a few miles south of town. Before leaving on the trip, I had checked several Airport Directories, all of them listed the airport as attended until dusk. As we crossed the Mississippi River into Illinois, I could see the airport. There was not a single airplane on the ground, the gas pumps sat out in the middle of an open field, and the place looked deserted. I landed first, shut down, climbed out. As Bill taxied in, I signaled him not to shut off his engine. I did not want to have a session with set of hot magnetos even though his engine had not given us any real trouble since Ashland.

While Bill waited, I walked the 300 yards to a house in a grove of trees. There was nobody there, even though the doors were open and a radio was playing. I looked around the garage, but did not see anything that would have helped us. Back at the Chief, Bill and I discussed the situation. We had two choices: Dubuque, Iowa was 25 miles away, but had a control tower and was somewhat off course. Freeport, Illinois was 30 miles away, closer to the course, and only 60 miles from Madison. The flight from Washington to Savanna had taken an hour and 20 minutes. Both Bill and I had plenty of fuel, so we decided to go to Freeport. Once out of the Mississippi River valley, there was a railroad going from Savanna directly to Freeport. All we had to do was follow it. The whole trip from Washington to Freeport took only an hour and 47 minutes of flying time.

A few years earlier, I had landed the Cub at Freeport on my way back from the National Biplane Fly-in at Bartlesville, Oklahoma. Unfortunately, I had arrived after the fixed base operation had closed. Being only 60 miles from home, I did not relish the thought of staying overnight. In that I had the EAA STC to use automobile gasoline in the Cub, I wondered where I could get some. Suddenly, I had a brilliant thought– Call AAA. I did and told the gasoline station attendant to bring 5 gallons of regular to the airport. When he arrived, he was surprised that I poured the gas into the Cub's tank. The $ 25 service charge was much less than a motel would have been.

Today, the airport office was still open. After getting fuel, Bill and I took off and headed the Cub and Chief north. As usual, Bill went ahead of me toward Madison and started to slowly pull away. It was a beautiful evening, warm, clear sky, and no wind. The setting sun gave everything a golden hue. Near the Illinois-Wisconsin border, I flew low past a farm house. A woman standing on the porch, in a blue dress with a white collar, waved to me. I waved back and did the usual wing waggle. From the expression on her face, she seemed pleased that I had responded to her.

While some 50 miles away, I was able to see the television towers at Madison. I also was able to keep Bill's Chief in view most of the trip, only losing sight of it as he approached Waunakee. I landed just as the sun set. The Cub was put in its hangar while it was still light.

Since our wives were not expecting us, Bill and I went to Pat O'Malley's restaurant, in Waunakee, for dinner. Pat flies a Piper Cherokee and has a hangar at the Waunakee airport. We gave him a quick rundown of our travels.

The 1,911 mile trip took 31 hours and 6 minutes of flying time at an average ground speed of 61 mph. Bill and I had made a gigantic loop through Illinois, Ohio, Kentucky, Virginia, West Virginia, Tennessee, North Carolina, South Carolina, Georgia, Tennessee, Missouri, Iowa and back home. We had flown over the wild terrain of the Appalachian and Great Smokey Mountains, where there was no place to land. Also over miles of pine forests again offering few opportunities for a good landing. The southeastern U.S. is well known for having nearly daily thunderstorms. Yet, we were never delayed by weather or even had to deviate from the planed flight path. Other than a bit of hot magneto trouble, both the Aeronca Chief and the Cub performed beautifully, but communication via my NARCO radio always was problematic. It was a good thing that Bill's radio, built from a kit, always worked perfectly.

Chapter 12
The Coeur D'Alene, Idaho Trip

After the Gastonia trip, Bill and I had decided that going west was a lot less nerve wracking than flying the mountains of West Virginia. Coeur D'Alene, Idaho was chosen as our destination as Bill had a college classmate who was teaching there. This city is northwest of the infamous Mullan Pass that Ann and I were unable to get past. The plan was to fly to Rapid City, South Dakota then to Billings, Montana. From there, we would follow Ann and my path to Missoula, Montana and then go over Mullan Pass to Coeur D'Alene, Idaho.

When the day of departure arrived, the weather decided not to cooperate. Bill and I watched the weather maps for three additional days before the weather became flyable. We finally departed Waunakee mid-morning on Thursday June 1, 1989, refueled the airplanes in La Crosse and headed for Mankato, Minnesota. Approaching St.Charles, Minnesota, I noticed that the turkey farms of the past were gone. A few miles before Mankato I flew over a large field that had not been planted. There was a small weatherbeaten farm house sitting all alone in the middle of the field. The paint had disappeared long ago, the bare wood had weathered to a dark grey color, and there was no glass in any of the windows. The remnants of road leading to the house had green grass growing in the tire tracks---another victim of farm consolidation. The leg to Mankato took two hours 25 minutes of flying time and that was stretching things a bit fuel-wise for the Cub.

After Mankato, the next stop was Marshall, Minnesota--95 miles away. Due to a headwind, our ground speed was only 48 mph and it dimmed our enthusiasm for plodding westward. As it was getting late in the day, Bill and I stayed over night there. From Marshall, it was 85 miles to Brookings, South Dakota whose University buildings can be seen from many miles away. I had flown past Brookings many times, usually in bad weather, but this was one of the few times that I would land here.

After refueling in Brookings, Bill and I flew to Huron, South Dakota where the flight path joined the "northern" route to Denver. Then, we headed across the rolling prairie toward Pierre, South Dakota. In this part of the world, I really enjoy getting on the deck and contour flying the hills and valleys. About two thirds of the way to Pierre, I was flying along and saw a big, radial engine, low wing, cropduster (called a Snow), pop up out from behind a hill, do a wingover and disappear going south. In a few minutes, it would pop up about a mile to the south, do a wingover and disappear going back to the north. I concluded it was spraying crops down in a valley and the best place to cross was halfway between the two wingovers. As Bill and I came over the edge of the hill, there was a broad valley behind it. I could see a flagman standing at the end of a field. By now it was too late to do anything, but fly right across the field at 300feet. I hoped we would not get in the cropduster's way. I saw the Snow pass underneath us with lots of room to spare, but I was dubious as to what the pilot thought of our overflight.

In the 95-degree temperature, I noticed that Bill would catch up to me more often than normal. This perplexed me as usually I would be in less wind than at the 3-400foot altitude he normally flew. On the ground at Pierre, I inquired about his higher ground speed. He said that I was flying too low and was getting into the superheated air near the ground. To stay airborne, it was necessary for the Cub to be at a higher angle of attack and consequently it had a lower ground speed. Later I tested out his theory and he was right.

A few minutes after we landed at Pierre, the Snow crop-duster also landed and taxied up to the gas pumps. Bill and I made ourselves as inconspicuous as possible. We were not sure if the pilot of the Snow was going to hassle us or not. Thirty minutes later, the flagman (a young woman) came over to us and she asked,

"Did you two fly over?"

I replied:"Yes."

"I saw you."

Bill and I breathed a sigh of relief.

There was a pick-up truck in the hangar with an A frame in the truck's bed. A Pratt and Whitney 450hp radial engine was hanging from it. Bill asked the mechanic about the truck. He said a Snow crop-duster had blown an engine and the pilot had landed in a field 25 miles away. It was easier to do an engine change in the field than it was to get the plane back to the airport. They were on their way out to the site to do the engine switch. We were impressed by this, but the mechanic said it was a common event.

Looking to the west, Bill and I could see a weather front and its line of small thunderstorms. In Pierre, all that was happening was a fairly strong wind. We had to wait a couple of hours for the wind to drop to an acceptable level and for the thunderstorms to dissipate. By the time we climbed back into our planes it was late afternoon. The destination was Rapid City with a gas stop in Philip.

The 70 miles to Philip took one hour and ten minutes of flying time--yielding a ground speed of 60 mph. The Philip airport is four miles east of town and is just to the north of a small river lined with cottonwood trees. This airport is one of the shabbiest I have ever seen. While flying the traffic pattern, I could see that one runway was completely overgrown with sagebrush and was unusable. The other runway was asphalt with large sandy areas showing through. It did not seem to be in very good shape. Despite this, neither Bill nor I had any trouble landing. Each time I land here a different fellow is the FBO. The new one always looks lower on the economic scale than the previous one.

Nailed to the office wall was a J-3 front wing strut with a huge hole through it. I asked the FBO about the hole. He said that he and another fellow had been hunting coyotes from the air. The guy sitting in the rear seat was tracking a coyote, when he fired his high-powered rifle he hit the wing strut.

Bill and I refueled the planes at Philip, taxied out, took off, and set up a due westerly heading. About halfway to Rapid City, we could see the buildings at Wall Drug and some of the Badlands several miles to the south. A few minutes later, we crossed Interstate 90 and a railroad track. Approaching Rapid City, Bill did the communicating with the control tower reporting in as a flight of two. By the time we landed, it was late in the day, so we decided to RON here.

The line boy had us taxi down a narrow alley between two hangars. Then, the planes were tied down behind the buildings. While we were doing this, a fellow from California and his girlfriend landed in a Beechcraft Bonanza. They came over and talked to us for a while. The fellow said they were going to get a rental car and would give us a ride into town. They used the car to move a large foot locker from the Bonanza to the office where they left it. In the mean time, Bill called a downtown motel and reserved a room. That evening, in the dining room of the Hiriam Johnson hotel, we saw the couple again, but they only gave us a cursory glance. Now, I started to wonder as to the contents of the foot locker.

The next morning(Saturday), Bill and I took the airline limo from the motel back to the airport. While we were packing up the planes, a fellow in his early 60's introduced himself as Bernie Havelar, a retired USAF Colonel. In the course of the conversation, he said he owned a Waco F-2. He asked if we would like to see it. The plane was in a row of hangars behind where we were parked. The biplane was a masterpiece in red with black and gold trim. It once had been the test bed airplane for Continental Motors. He had restored it with their logo on the side of the fuselage. In the hangar, he also had the most beautiful C-85 engine I had ever seen. If I had owned it, I would have been content just to sit and look at it. It looked better than when it was new. After we returned home, I asked Otis Lokken about Bernie. Otis said that Bernie had talked to the Stoughton boys a number of times at the Waco Fly-in. They had written him off as an informed bystander, until one year he showed up with his beautiful F-2. His stock went up dramatically. Bill and I had spent the better part of an hour with Bernie. In order to get back into the air, we reluctantly had to break off the conversation.

Taking off to the northwest, I could see B-52's on the ramp at the Ellsworth AFB. The Rapid City tower called me and told me I was too close to Ellsworth. I quickly made a steep turn to the left, then I headed for Interstate 90 that goes up a valley on the east edge of the Black Hills to Spearfish.

As I rolled out on my landing at Spearfish, a Cessna 172 taxied out to where the taxiway and the runway meet. I knew Bill was landing behind me, so I needed to get off the runway. I was taxiing to the 172's left when I heard a big clunk under the floorboards. I did not have a clue as to what caused it. I shut down the engine, climbed out, and saw that I had hit a taxi light with the Cub's right shock strut. Another light was bent over under the Cub's belly. I was lucky that the propellor had not hit the lights. These lights were right on the edge of the asphalt. Since my attention was on the 172's wingtip, I never saw them. When I paid the gas bill, the lineboy was told that I had broken a couple of taxi lights. I gave him my name and address. I fully expected a bill for about $100, but the old gal who runs the airport never sent me one.

After takeoff, the Cub and Chief were pointed towards Gillette, Wyoming. As we came close to Gillette, it became apparent the airport was completely different from what it had been on my last visit. Now, there were: a tall concrete control tower, several very large hangars, and a 7,000foot long concrete runway. My radio was working, so I called the tower for landing clearance and touched down with Bill right behind me.

As there was no restaurant at the airport, I inquired about getting something to eat. The lineboy gave us the keys to the airport courtesy car, so we could go the three miles into town for lunch. This car was a nightmare, nearly nonexistent brakes and worse steering. When we returned, I mentioned the brakes to the lineboy, he laughed and admitted they were bad. I felt a lot safer in the Cub than in that car.

The next leg was from Gillette to Sheridan, Wyoming on a northerly heading. Landing at Sheridan, I noticed that the quaint 1930's white stucco terminal building had been replaced by a new metal building having no charm at all. After refueling the planes, Bill and I took off under a high, dark grey overcast, and headed north. To the west, the Big Horn mountains loomed over us. Far off to the east, the sun, shining through holes in the cloud deck, was giving patches of the prairie a golden hue.

About 10 miles north of Sheridan, scattered rain showers started falling from the high overcast. Each of the showers was only a few hundred yards across, but there were quite of few of them. The falling rain formed light grey, almost white, columns from the cloud base to the ground. They looked like the pillars in a huge Gothic cathedral. To the east, they were back lit by the sunny fields near the horizon. Bill was off to my right working his way around and through the edges of the columns of rain. Against the several shades of grey, his plane looked like a small yellow bird trying to find its way out of a cathedral. As the Chief flew behind a rain column, it would disappear into the mist for a few minutes and then suddenly reappear.

In the meantime, I was working my way around the denser rain columns. Occasionally the Cub and I would fly through a rain shower's grey veil of light mist. Both Bill and I did this for the better part of 15 minutes, while keeping reasonably good track of each other. As we flew around and through the columns of rain, we slowly worked our way along the course line to the north. I had an overwhelming sensation that we were flying inside a gigantic Notre Dame. For me, the experience took on somewhat mystical overtones. I think it was because of the immense scale of the weather formations and the insignificance of our little aircraft against this background.

Ordinarily, we would have flown to Billings, Montana on this leg, but the course was plotted to Laurel,

Montana a few miles to the west of Billings. The plan was to turn west at Laurel and then go to Livingston, Montana. As the Cub and Chief proceeded along this route, thunderstorms started appearing in the west, but there were holes in the cloud deck where the sun shone through. Nearing Laurel, we could see that the far end of the valley leading to Livingston was filled with clouds.

At the Laurel airport, there was a single maintenance building and a few hangars. After landing there, Bill and I encountered a time warp. The fellow running the airport was a hippie right out of the 1960's. He had been a chopper pilot in Vietnam and made his living doing contract flying for the Forest Service with his own Bell 47 helicopter. There was a Cessna 172 hauling parachute jumpers, who also were right out of Madison in the 1960's. The only fuel he had was high lead 100-octane that is very hard on low horsepower engines such as ours.

After a conference about the weather to the west, Bill and I decided we would not be able to reach Coeur D'Alene and would make Billings the western terminus of our trip. I think both of us were getting a bit saddle sore from three days of flying. By now, my radio was in it's receive only mode, so Bill would have to lead and call in a flight of two at Billings. This would be the last time we would be able to fly into the Billings airport. On the first of July, a traffic control area (TCA) was going into effect and mode C electronics would be required.

Bill took off first from Laurel. Before I could takeoff, the jumper's Cessna would have to leave the immediate area. When I did get airborne, I could not see Bill's Chief anywhere. I flew the Cub in the general direction of Billings while making wide sweeping S turns, but still could not see it. After 10 minutes of this I was concerned that Bill would reach Billings before me. Finally, I saw the Chief off to the northwest and joined up with it. Later, Bill told me that he had the Cub in sight the whole time. After landing, we taxied downhill to the east end of the field, parked the planes on the grass, and tied them down. The 1,130 miles to Billings had taken 18 hours, 13 minutes of flying time at an average ground speed of 62mph.

Parked behind an airport building was a Vietnam era B-26K, with wing tip tanks. Talking to one of the locals, we learned that the FBO owned five B-26's. He used them as water bombers for fighting forest fires.

Bill and I walked downhill on the airport road to the nearest motel and checked in. From there, we walked to the center of the city. In contrast to the small towns in this part of the world, Billings had many new buildings, wide streets, and looked prosperous. At dinner, Bill suggested we visit the Custer National Monument at the scene of the Battle of the Little Big Horn. I was not too enthusiastic about the idea as what appeared to be bad weather was lurking off to the west.

Bill prevailed, so the next day (Saturday, June fourth) we rented a car and drove to the site. The area had been nicely preserved and I was glad that we made the 60 mile trip. I was able to get a good sense of what had happened during the battle. It was one of the few battles where the U.S. Army was on the top of the ridge rather than at the bottom.

In mid-afternoon, we returned to Billings and turned in the car. The night before, Bill and I had re-planed the trip to make a big loop through North Dakota, northern Minnesota and then southeast into Wisconsin. Retracing our path through Rapid City would have just been a repeat of country I had seen many times.

While we were loading our luggage in the planes, a local fellow, who owned an Aeronca Champ, came along and offered to prop the Cub. I am very hesitant to let someone else do this, but he was very insistent. He flipped the prop as if the engine had impulse magnetos. The engine did not start. I had the feeling that I was in for trouble. It took him nearly 30 minutes to get the engine running--no more of that.

The segment to Miles City, Montana was 132 miles long. The flight path was along the south side of the

Yellowstone River that made lazy S turns as it ran to the northeast out of Billings. On the south side of the river, there were many green fields in the countryside, but only a few trees could be seen. To the north, where the prairie met the river, small cliffs lined the river. The skies were clear, there was a headwind, but no turbulence.

A few miles outside of Billings, Bill and I flew past an auto junk yard. In the middle of it, was an old moving van with Baker Transfer still visible on its side. My late uncle, Everett Baker, had owned a moving and storage company in Billings. I was sure that the truck was one of his. About half way to Miles City, I checked the ground speed, 55 mph and it made me anxious as there still was a long way to go. Thirty-five miles before Miles City there was a small airport. I seriously considered landing there for fuel. As we flew over the field, there was no sign of life on the ground. Now, I was committed to reaching Miles City.

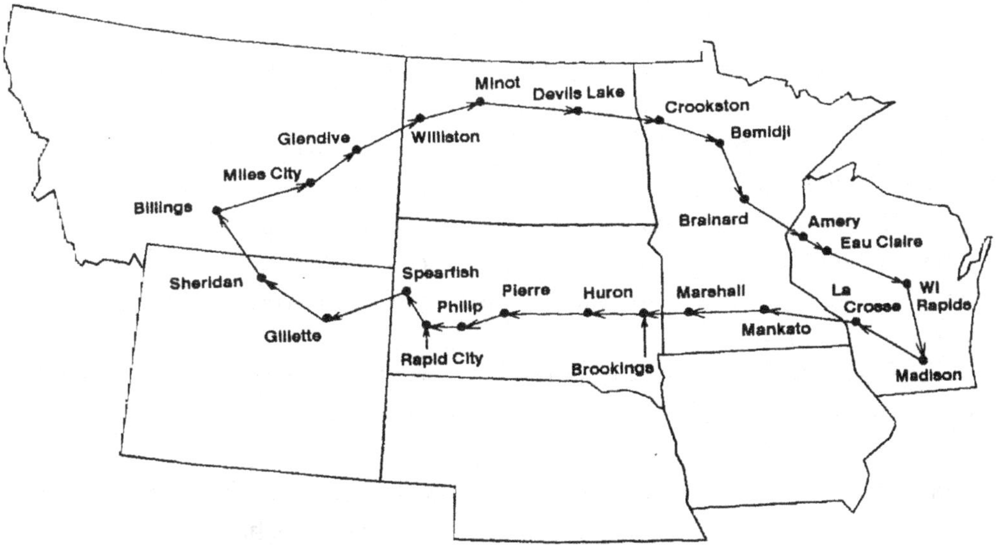

Modified Trip Map

After two hours and 30 minutes in the air the Cub and Chief landed at Miles City. The airport office was closed, but the FSS was open. The woman working there suggested that Bill and I stay downtown at the Olive Hotel. It turned out to be a good recommendation. The hotel was a typical three story brick building built in the 1890's whose interior had been modernized. Bill and I had excellent meals in its small dining room. Afterwards, we walked around town and through a large park bordering the Yellowstone River. All-in-all it was a nice small town and one of the few that we had time to enjoy.

The next morning (Monday), the hotel owner gave us a ride back to the airport. A pilot of a Mooney 21 was waiting at the gas pit to refuel his plane. Bill and I rolled our planes in behind his. Upon talking to the fellow, I learned that he knew my late Uncle. He said that Everett had been a very prominent person in Billings and he knew him well--small world.

The next leg followed the Yellowstone River northeast to Glendive, Montana. JoAnn's brother-in-law, Frank Fager, was from Glendive. All of our married life, I had heard about the city. Now I was going to see it for myself. Leaving Miles City, Bill and I were still flying in a wide river valley, but off to both sides the landscape started to change from rolling hills to badlands. The ground was eroded into numerous small pointed hills having red and brown horizontal strata. It was like a miniature of the region near Sedona, Arizona. Closer to Glendive, the river valley narrowed with cliffs on both sides and the countryside was barren wind blown sand. Just to the east of town, there was a rather tall, tree covered bluff.

The Glendive airport was 5 miles to the northwest of town. Except for a fellow running a crop-dusting service, there was not much at the airport. He had a dog and a cat that were attracted to our airplanes. The dog kept marking the tires and the cat insisted upon parking itself in front of one of the Cub's wheels. When it came time to start the engines, the cat refused to move. It took a few minutes to get the cat out of harms way. After takeoff, I circled town a couple of times. Two sets of railroad tracks ran through town. In its prime, Glendive was where the Northern Pacific Railroad crew changes were made. It also was a maintenance center, but the large roundhouse did not appear to be in use. I took several aerial photos of the town. I knew that Frank Fager would enjoy seeing them. Unfortunately, none of the pictures included his homestead.

From Glendive, Bill and I followed the Yellowstone River northeast to Williston, North Dakota where the Yellowstone joins the Missouri River. At Williston, both planes were refueled. After takeoff, the heading to Minot, North Dakota was due east. The first part of the 110 mile flight took us alongside of the upper end of a large reservoir called Lake Sakakawea. It was created by the Garrison dam on the Missouri River some 75 miles to the southeast. The land on the north side of the lake, nearest to us, was covered with scrub trees. On both sides of the lake there were few signs of civilization.

Forty miles past Williston, the Cub and Chief slowly went away from the lake into ranch land. Our bad luck on long flight segments was still holding. A ground speed computation revealed we were only making 45 mph. After leaving the lakeshore, there were only a few roads to serve as navigation checkpoints, but the city of Minot could be seen from quite a distance.

We also were only 50 miles from the Canadian border. In Minnesota, the natives say that the winter's cold, arctic air blows in from Canada and there was nothing in North Dakota to stop it. Flying along this segment Bill and I had proof they were absolutely correct.

The Minot airport was north of town and had a control tower. At 5 miles out, Bill called in and reported a flight of two on a left base for runway 32. A few minutes later, I called in on a right base for 32, but the tower had some difficulty understanding my transmission. The tower operator called Bill and asked him who was on a right base. Bill replied it was the Cub, but it was on a left base. I never will learn left from right.

After being refueled in Minot, the Cub and Chief continued due east towards the city of Devils Lake. I thought that I had seen flat land in Illinois and Indiana on my trips east, but that country was hilly compared to North Dakota. There was not a single contour anywhere along the line of flight. Surprisingly there were quit a few small lakes. A few miles from the city, I could see flocks of birds floating on a large body of water. It was the lake for which the city was named. The wind had let up a bit as we made an average ground speed of 60 mph on this leg.

By now it was late in the day, so we decided to remain over night at Devils Lake. Since the airport was a couple of miles out of town, we took a taxi to the nearest motel. This was another town with no redeeming features. I have no recollection of where we stayed or had dinner.

Although we could see the airport from the motel, the next morning(Tuesday) we took a cab back to the airport. Bill and I loaded up our planes and took off at 8:00 A.M. The Cub and Chief were pointed slightly to the southeast towards Crookston, Minnesota some 106 miles away. The line of flight crossed Interstate 94 several times. It also took us through the control zones of both the Grand Forks Air Force Base with its B-52's and the Grand Forks International Airport. Bill and I avoided both of these areas by flying a few miles further to the south. Just in case there was any traffic, we monitored their radio frequencies. The University of North Dakota has a very large pilot training program that is partially funded by Northwest Airlines. Normally, there should have been some student pilot traffic, but today things were quiet.

After the run down buildings we had seen at most of the small airports, the Crookston airport terminal building was a pleasant surprise. It was a new brick building with lots of glass and a nicely decorated interior. The fellow on duty said the State of Minnesota had been financing airport modernization. Crookston was one of the lucky airports to receive funding.

Several Piper Pawnee crop-dusters were taxiing up to a hangar and refilling their pesticide tanks. After turning around, they took off in whatever direction they needed to go. Consequently, Bill and I had to keep a sharp lookout when we taxied out to the runway and took off.

As the Cub and Chief proceeded toward Bemidji, Minnesota, the countryside changed from crop land to mixed farms and pine forests with many lakes. Bemidji also had a new terminal building and there were numerous airplanes on the ramp. Bill and I had lunch at the airport terminal (a rare treat) before heading further southeast to Brainard. This leg took us over the heart of the Minnesota vacation country. Bill and I flew right along the southwest shoreline of Leech Lake that is a very large lake.

Twenty miles before Brainard, Bill and I flew over the southwestern edge of Whitefish Lake. In the early 1940's, my mother, brother John, sister Donna, and I had spent two summers in a cottage on a small lake connected to Whitefish Lake. The three of us children would row a boat through a narrow channel to the larger lake in order to play on the sand bluffs.

Today, from the Cub, I could see those white sand bluffs in the distance. I was tempted to divert from course to see if the cottage was still there. However, Bill and I had not discussed doing this before hand. With my marginal radio there was no way I could easily explain to him why I was diverting some 10 miles off course to look at what was probably torn down long ago. Consequently, I passed up an opportunity to revisit a very pleasant past.

During those summers, I had traveled by car through this lake country. I always had the impression that the lakes were surrounded by dense pine forests. Yet from the air, it was apparent that the woods extended only a few hundred yards back from the lakes--another illusion destroyed. Bill and I landed at Brainard that was another of those relatively busy tourist oriented airports.

After refueling the planes, we took off again and headed for Amery, Wisconsin. Twenty-five miles past Brainard, Bill and I flew along the southwest shore of Mille Lacs Lake that is another popular vacation destination. Along the shore road were the typical white painted wooden tourist gas stations, grocery stores, and souvenir shops. Since I was flying quite low, I was able to read the signs on the fronts of the buildings.

Some 82 miles past Brainard, we came to the St. Croix River valley northeast of the Minneapolis metropolitan area. There are state parks on both sides of the river. The river itself is quite wide and is designated as a national scenic waterway. The designation was a good one as the river wends its way through some beautiful densely wooded forests with very tall pine trees. At this point, our course line followed the river for some 15 miles. Bill called me on the radio to warn me that this area was a no low flying area. Although the maps gave no indication of this being the case. Just to be on the safe side, I climbed up to 600feet AGL, so the Cub's engine would not be heard from the ground.

Shortly after departing the river valley, we reached Amery, Wisconsin whose airport was down in a depression. While we were refueling, a major fire broke out a few miles south of the field. A tall column of black smoke rose up some 1,000feet and then hung over the countryside. One of the Amery airport locals went in his car to investigate the fire, but did not return before Bill and I left. It was a short hop (45 minutes) to Eau Claire where the planes were tied down for the night.

The next morning (Wednesday, June 7,1989), after repeating the long ride in the motel's van back to the airport, we loaded the planes, and took off. The Cub and Chief were headed southeast towards Wisconsin Rapids, Wisconsin--83 miles away. The line of flight took us over mostly wooded land, past a 1,000foot tall television tower, and the Nellsville airport that sat on top of a hill. We were in a Fort McCoy MOA for part of the way. This required our paying some attention to the sky around us as Air National Guard A-10's could be flying in the area.

In the past, Bill and I had landed at Wisconsin Rapids a number of times, so were getting back into home territory. A few miles before the airport, I heard the pilot of an Aero Commander light twin call in on a five-mile final. I flew on for another few minutes and never heard another message from him. Assuming he had landed, I set up for a right-hand base leg to the runway. Just as I started turning onto final approach, the Aero Commander passed right in front of me and a hundred or so feet below me. It really shocked me. At its ground speed, it should have covered the 5 miles a long time ago. I think the airplane was really about 10 miles out when the pilot called in. He thought that nobody else was in the air (you are never alone!)

By the time Bill and I landed at Wisconsin Rapids, it had become quite warm. The wind was from the southwest meaning that we would use runway 20 for takeoff. I did not like this runway as there are tall trees across its end, a golf course in the trees, and some tall buildings in the city beyond that. Just as I thought, as soon as the Cub broke ground, it started acting sluggish and gave little or no sign of wanting to climb. I started going through the usual struggling for altitude drill. The Cub barely cleared the first trees and then went skimming across the tree tops. Finally, the Cub decided it had my attention for the day and started to climb. As usual, Bill had no trouble and climbed right out. A while later, I could see the Baraboo Bluffs through the thin haze. I thought that Bill would start his "off to the barn" act, but he dutifully kept up his slow S turning routine all the way back to Waunakee.

This circular 2,263 mile trip was the longest of the Vogelsang trips. It took 36 hours 50 minutes of flying time at an average ground speed of 61 mph. Bill and I had at least some headwind nearly every inch of the way. The weather was very benign for the whole trip. A large part of it was flown under blue skies with tremendous visibility. The advantage of going west rather than east. The only regret was that Bill and I did not make it to Coeur D'Alene, but that would have added another three days to an already seven day trip. I'm not sure that either of us would have wanted to strap ourselves into the Cub or the Chief for that much longer.

The “Coeur D'Alene” trip also turned out to be the last of the Vogelsang trips. The following summer, Bill purchased a run-out Piper Tri-Pacer in Bloomington, Illinois. He and I brought it back to Madison on a trailer at night in a blinding rainstorm. It rained so hard that I could only see a hundred feet of the road most of the trip, but we made it. Since then, Bill has been engaged in restoring the plane to flying condition. Unfortunately, the Tri-Pacer is a good 30 mph faster than the Cub, so it would be difficult to take both planes on another Vogelsang trip.

Chapter 13
Sentimental Journey to Lock Haven

The manufacture of Piper J-3's began in 1937 at Lock Haven, Pennsylvania and ended in the spring of 1947. In the mid 1980's, a number of former Piper employees decided to celebrate the 50th anniversary of the first J-3. To accomplish this, a Piper Cub Fly-in was scheduled for the middle of July 1987 at the Lock Haven airport. Being a died-in-the-wool Cub fan, I could not pass up such an event. Unfortunately, I had to teach summer school and had only four days in which to get to Lock Haven and back. Because of the time limitations, JoAnn and I drove there. Only 39 airplanes were in attendance, but not all of them were J-3's. Despite this, it was a nice low-keyed event. The people I talked with were quite interesting. The next year, much to my dismay, the fly-in dates again conflicted with summer school, so I was unable to attend.

At the November 1988 banquet of the Midwest Antique Airplane Club (MAAC), I had talked to Bill Coleman about flying our Cubs to Lock Haven. The Fly-in, now called "Sentimental Journey," was scheduled to be held in late August,1989, which was after summer school. Bill was enthusiastic about the idea. He indicated that his father would accompany him. I had known Bill since he was about 20 years old. He and his wife went on their honeymoon in his Cub. They had camped under the Cub's wing the whole time. Now, they owned his and her's Cubs. Bill has a grass runway and a hangar across the road from his father's farm house. I knew Bill was my kind of a Cub pilot and would be a good man to fly with to Lock Haven.

In July of 1989, I laid out a 711 mile route to Lock Haven that was as direct as possible. It also maximized the number of airports having 80-octane gasoline. In contrast to the Gastonia trip, it was only a two-map (Chicago and Detroit) trip.

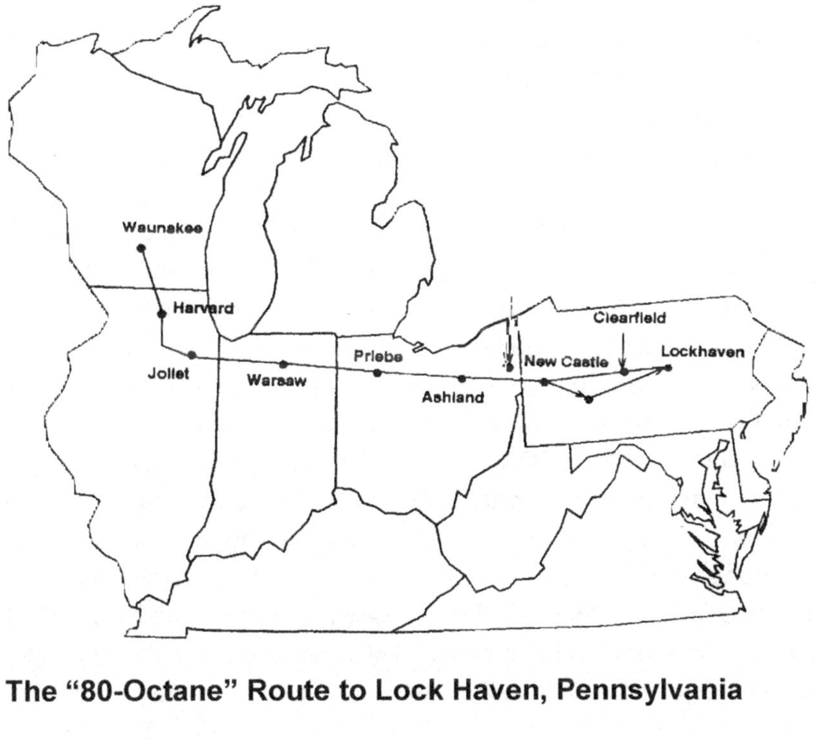

The "80-Octane" Route to Lock Haven, Pennsylvania

Sentimental Journey was to be held from Thursday, August 23, through Sunday, the 26'th. Bill and I planned to depart Waunakee on Wednesday the 22'nd and arrive at Lock Haven on Thursday. Early Wednesday morning the weather was low clouds, rain, and fog. I called Bill and we agreed to try again the next morning. The Thursday morning weather still was not flyable, but I went to the airport with hopes that it would improve. By 8:00 A.M., the clouds had lifted to about 300feet and the visibility was improving. Twenty minutes later, I heard the hum of a Cub's engine and soon thereafter Bill and his father landed.

His father was a short, slim 72-year-old Irishman. Many years ago, he had gone to Oklahoma with a local pilot to ferry a J-3 back to Portage, Wisconsin. At one of the airports, they were taxiing when the wind flipped the Cub onto its back. A steel fence post came through the roof of the plane right behind his head. Yet, this experience had not dampened his enthusiasm for Cub flying.

It was agreed that I would be the lead navigator and Bill would follow me. The initial leg went around the west side of Madison's cluster of television towers and then to Harvard, Illinois -- 72 miles away. Around 9:00 A.M., the ceiling was high enough for us to takeoff. Once both Cubs were off the ground the clouds were lower than we thought and at the northwest edge of Madison they touched the ground. To the east, there were low scud clouds, but I could see the state capitol building and Lake Monona beyond it. Flying under the clouds, I would occasionally enter a piece of scud. As I flew out of one of these clouds, a Cessna 172 crossed 100 yards in front of me at exactly my altitude (you are never alone). I was shocked to see him and am sure he never saw me.

By the time the Cub reached the southern limits of the Madison suburbs, the ceiling had gone up some. The open areas between the scud were more numerous, but it was still very hazy. I couldn't see Bill's Cub, so I landed at the Stoughton airport. He came in right behind me. He said that, when my Cub disappeared into the scud, he had some trouble keeping track of it. By flying straight ahead, he would come out of the scud just behind me. We decided that the worst of the weather was associated with Madison's lakes and that it would improve to the south.

Bill and I took off and forged ahead to Dacy Airport at Harvard, Illinois. This is a very interesting airport with many antique airplanes based here. All three of the runways are grass and there is a single row of metal hangars. At the north end of the airport there is a large hangar with an airport office attached to its side. The Dacy family is in the business of restoring Stearmans. Their hangars usually contain three or four of them.

The next stop was Joliet, Illinois--75 miles away. Initially, Bill and I could follow the highway that goes directly from Harvard to Marengo. Just past Marengo is a private airstrip. Where a DeHavilland Tiger Moth biplane and a DeHavilland Hornet Moth are based. The town of Plano is 60 miles due south of Harvard. It has a low, but rather large water tower that makes a good checkpoint. South of Plano, the course line crosses a railroad track where a turn toward Joliet was made. Bill and I landed there at 11:30 A.M. and refueled the Cubs.

The next fuel stop was Warsaw, Indiana--122 miles away. About ten minutes out of Joliet, Bill and I flew past the Howell-New Lenox airport and in a few minutes more the Frankfort airport. Both airports are just on the edge of the Chicago Terminal Control Area(TCA). This part of the flight segment concerned me as most of it is over contiguous cities. Other than at these airports, there are few places to land. At Crown Point, Indiana, some 40 miles east of Joliet, the last of the Chicago suburbs were left behind. Now the countryside was flat farmland with occasional groves of trees. Most of these groves were in the form of a well defined rectangle. It seemed as if each farm had a grove. A compass heading of 95 degrees would hold for the next 250 miles.

Because this was my first flight to Lock Haven, I was paying particular attention to staying exactly on course and making mental notes of the countryside. If the weather sours on the return trip, this knowledge would come in handy.

To the east of Crown Point, there is a lake, shaped much like the map of Wisconsin. It is bracketed on its north and south ends by tall television towers. The course line went right between the two towers. Forty-two miles west of Warsaw, Indiana we flew just north of the Starke County airport where no activity could be seen. There was a white VOR building about a half mile southeast of the airport. Ten miles before Warsaw, at the village of Bourbon, the Cubs crossed the railroad track and four lane highway that goes from Plymouth to Warsaw.

At Warsaw, the two Cubs were topped off and we used the airport car to drive to a nearby restaurant, named Grandpa's. After lunch, we took off for the Priebe airport located 12 miles north of Findley, Ohio. On this leg, we passed north of Fort Wayne, Indiana whose tall downtown buildings stood out against the horizon. The next checkpoint was a rectangular lake north of the town of Grabill. A few minutes after the lake, we crossed the upper end of a reservoir on the Augliaze River.

West of the Priebe airport, a 1,094foot tall television tower was exactly on our course line. The tower could be seen from many miles away, so it was not a threat. The only runway at Priebe is a real jewel, 2,000feet long asphalt, running from the southwest to the northeast. At the southwest end are a power line and a telephone line with the latter right at the end of the runway. At the halfway point, the runway goes through a gap in a woods, and ends at an open field. At the far end of this field is a large landfill with flocks of sea gulls milling around looking for something to eat. As I landed to the southwest, deep water filled ruts in the ground could be seen on both sides of the narrow asphalt runway. It looked as if heavy equipment had driven through on a muddy day.

The FBO was a well-known P-51 owner. How he flew in and out of here in his P-51, I will never understand. The asphalt runway was just barely wider than the span of the Cub's landing gear. I talked to one of his mechanics about landing the P-51. He said that the FBO had grown up at the field and had never had any trouble landing his P-51 here. In his shop, they were restoring a WW II F4U Corsair. The unrestored parts looked very rough.

As we proceeded east, the sun was getting very low in the western sky, so camping overnight at the Wayne County, Ohio airport seemed problematic. Bill did not have a radio, so I could not confer with him. Knowing that Bill would follow me, I decided to land at Ashland, Ohio.

By the time the two Cubs approached Ashland, Ohio, there was not a lot of flying time left in the day. From a few miles out, I could see the airport sitting on the top of a high hill to the east of the city. I landed to the south just as a National Guard helicopter departed the runway. Bill and his father came in right behind me. The airport manager, named Johnson, seemed pleased to see us. He said we were welcome to camp overnight. Bill and I parked the Cubs on the grass between the hangars and the runway and tied them down. We also set up our tents under the wings, which might shield them somewhat from the evening dew.

Mr. Johnson loaned us his car, so we could drive the four miles down the hill to a restaurant near the edge of town. At dinner, Bill's dad related tales of his great-grandfather who, in the mid- 1800's, had ridden rafts of white pine logs down the Wisconsin River into the Mississippi River and then to New Orleans. The astounding part of the tale was that the loggers then walked the thousand or so miles back to Wisconsin with the proceeds in their money belts.

From previous airplane camping trips, I knew that getting breakfast could be a problem. Consequently, Bill and I purchased some rolls and orange juice in the grocery store across the street from the restaurant. Mr. Johnson, thoughtfully, left the airport office open for the night. The three of us awoke the next day to a clear, warm, and calm day. The tents were put away and the breakfast goodies consumed. We took off before 8:00 A.M. and headed southeast. The main part of the city of Wooster is on the east side of a deep valley and follows a highway up a hill to the north. Our two Cubs flew directly over a large high school perched high above the city.

To the north, I could see the Wayne County airport. It turned out to be a long way from Wooster. I was glad that we had stopped at Ashland.

On our way to New Castle, Pennsylvania it was necessary to avoid the Akron-Canton Airport Radar Service Area (ARSA). To accomplish this, the course line went around the south side of Canton, Ohio to a VOR building. This building is very difficult to find as it is perched amidst the woods atop a high hill. At this point, the line of flight turned to the northeast and some 30 miles further went over the city of Salem.

Once over the Youngstown-Elser airport, Bill and I took up an almost due east heading. Just west of the New Castle airport there is a deep valley with a river at its bottom and double railroad tracks on both sides of it. To the east of the airport is another deep valley with power lines, an Interstate highway, and a river at the bottom. Thus, both ends of the southwest-northeast main runway have a 500foot drop off. Despite this, we had no problems landing to the southwest on the somewhat uphill runway.

From New Castle, it was on to Punxsutawany, which is impossible to pronounce. At about the halfway point, the two Cubs crossed the narrow, but deep valley of the Allegheny River. A string of hamlets are situated in the bottom of the valley and along the banks of the river. Four miles south of the course line, there is a power plant on the river flats. Its 1,010 foot tall smokestack reaches high above the tops of the valley's bluffs. It made an excellent checkpoint even though it was not directly on our course line. By now, the land below us had become quite hilly with a number of north-south valleys. Typically each valley had a creek and a railroad track at the bottom with villages strung along the railroad track. There also were small farms in the valleys. Being a locomotive engineer by profession, Bill could not resist following the railroad tracks that wended their way across the hills and valleys. He was particularly interested in tunnels and railroad trestles. I would look back and see him heading off-course to inspect them. Eventually, he would pick up the course line again and follow me.

Although, the airport directories listed the Punxsutawany airport as attended, it was not. This is a recurring problem as most directories are updated only yearly. In addition, changes to the airports fuel status are not always reported. Now, Bill and I had a fuel availability problem. Fortunately, we found a woman and her husband painting a Cessna 172 in one of the hangars. She sold us enough fuel to fill our Cub's tanks. The next stop was Lock Haven.

Twenty miles to the east of Punxsutawany there is a high ridge called Clearfield Mountain whose top is at 2,410feet. There are several radar domes and a number of tall radio towers on top of this ridge. Our heading of 130 degrees took us just beyond the south end of the ridge and into a wide valley going to the northeast. Bill and I followed this valley to Clearfield where there is an open pit coal mine and a large power plant to the northeast of the airport. The plant's tall smokestack appears to loom over the airport, but actually is a mile or so away. We overflew the unattended Clearfield airport and took up a 92 degree heading to Lock Haven.

The West Branch of the Susquehanna River was followed for a while. This river is in a deep valley that is very narrow, has many tight bends, and whose sides are wooded from top to bottom. A railroad track clings to the river bank. The river itself is a rusty orange color while its banks displayed hues of reds and yellows. From the air, it looks like a classic case of water pollution caused by the coal strip-mines.

One thing that did impress me on this leg was the reclaimed strip-mines. The State of Pennsylvania has done a superb job of reclamation. All along the route, the mines had been filled in, the original contours restored, and trees planted. It was a dramatic contrast to what Ann and I had seen a decade earlier. At that time, many parts of eastern Ohio and West Virginia had been utterly devastated by strip mining. Those states could take a lesson from Pennsylvania.

Thirty miles east of Clearfield, Bill and I ran out of what few signs of civilization there were. Below us it was solid forest in all directions. Other than the tree tops, there was no place to land. As the pucker factor was getting high, I climbed the Cub up to 600feet AGL to be able to see a wider span of country. The higher altitude really did not make a lot of difference as it was still solid forest underneath. The weather was beautiful, so there was no problem navigating. Bill was flying higher than I was. He took a photograph of my Cub in which it appears as a tiny yellow airplane against a solid background of dark green tree tops.

The two Cubs crossed the deep valley of the West Branch of the Susquehanna River several times. After the last crossing, the river went off to the northeast and out of sight. A few miles further on, we came to another narrow deep valley with a creek at its bottom. There was a hamlet, Orviston, tucked alongside the creek. The valley is so narrow that this hamlet can not be seen unless you are directly above it.

Just before Lock Haven, there is a last high ridge to cross. Then, you are back in the valley of the Susquehanna River. It was a thrill to come over the ridge and see the old Piper factory far below. It is a several city blocks long low red brick building located at the southwest corner of the airport. This former silk mill is where the Cubs and many other Piper airplanes were manufactured. At the east end of the factory there is a separate rectangular modern appearing building with white walls and a flat roof. This was Piper's research and development building.

The top of the last ridge is at 1,810feet, while the Lock Haven airport is at 555feet. Having considerable altitude, Bill and I started our letdowns into a right-hand downwind leg for a landing to the west on the grass runway adjacent to the hard surfaced runway. At 2:06 P.M., my Cub touched down at its birthplace. Since we were a day late, Bill and I had to park our Cubs at the far west end of the parking area in almost the last row of airplanes. In contrast to my last trip to Lock Haven, there now were about 150 airplanes. Roughly 30 percent of them were J-3's. Apparently, the word had gotten out about the Sentimental Journey Fly-in.

After registering, the three of us walked around for a few hours. In the process, we said hello to Dick and Dan Knutson, Don and Steve Johnson, and John Vondran all of whom are Madison area pilots. After dinner, I took the van to the Holiday Inn, but Bill and his dad stayed behind to look around for a while longer.

One thing the organizers of the Fly-in have right is the transportation system. The Holiday Inn is the biggest motel in the area, but is 10 miles from the airport. The Fly-in organizers have several vans going to and from the motel on a regular schedule. If you need to go at an off-time, someone would take you there in a Buick or Ford car provided by local car dealers. The transportation system took the stress out of staying far away from the Fly-in site. Getting a motel room can be a problem as the World Junior Baseball Championships were held on the same weekend at Williamsport, Pennsylvania only 30 miles away. If you did not make motel reservations several months in advance, you ran the risk of having to camp the whole weekend. I like to camp, but in 90 degrees, 80 percent humidity, and no wind; an air-conditioned motel is a much needed luxury. The contingent from Iowa always camped on the north side of the airport next to the trees. Because they were there every year, this group was becoming a Lock Haven institution.

The Fly-in facilities were located in a grassy area on the south side of the runway. In the midst of a grove of trees, a pavilion with four long rows of picnic tables serves as the social center for the Fly-in. The pavilion is a popular place to sit in the shade. A conversation can be struck up with whoever is sitting nearby. I have met many very interesting people in this manner. Adjacent to the pavilion are three trailers whose proprietors sell food at very reasonable prices. One of them sells huge ice cream cones in a wide variety of flavors. Alongside the lower floor of an old barn, there is a small food stand, that is renown for its excellent breakfasts. During the day they have a grill going and serve hamburgers. At all of the food stands, large cups of lemonade and soft drinks are available that are greatly appreciated in the heat of an August afternoon.

Saturday morning was spent walking around looking at airplanes and talking to people. There was one freshly restored J-3 from Rubicon, Wisconsin. I had never heard of the town and upon talking to the young couple flying it, discovered they were Steve and Sharon Krog from Hartford, Wisconsin. They had a Cub of their own, but had flown this one to Lock Haven while the owner drove a motorhome. Steve and Sharon were quite active in the Midwest Antique Airplane Club, but it was odd that I had never met them back home. Dick Knutson introduced me to a woman who was based at the same Annapolis, Maryland airport that I had flown Cubs from in 1948. During WW II, she taught aircraft mechanics in the Army Air Corps. Currently, she flies a Piper Comanche.

One fellow, from Ft. Wayne, Indiana, had a display by his Cub describing his trip to Puerto Rico. He flew his Cub from Florida to the Bahamas. From there, he flew across 150 miles of open water to the Dominican Republic, and then on to Puerto Rico. If he had gone down, he was simply shark bait.

One Cub, of particular note, was one of the 50 J-3s powered by a Continental A-40 engine. It was the oldest J-3 still flying and was owned by a Federal Express pilot, who lives in Ludlow, Vermont. He also attends our Grassroots Fly-in, at Broadhead, Wisconsin, every other year. A few years later he sold this Cub to the Owls Head Museum in Maine. His understanding was that the Cub would be a permanent part of their aircraft collection. Much to his dismay, they quickly resold the airplane.

Saturday afternoon, I sat on my campstool next to the Cub and waited for people to wander by. I had several interesting discussions with old time Piper factory workers. One told me that Piper had a large building near Clearfield filled with parts for the fabric-covered Piper airplanes. When the company was taken over by Chris-Craft, some Harvard MBA decided they weren't making enough profit on these parts. He ordered that they be scrapped. The old fellow had tears in his eyes when he told of emptying the parts onto the parking lot to be hauled away by a scrap dealer. Today those parts would be worth their weight in gold. Not all of the parts were scrapped as, under the cover of darkness, many local pilots liberated as much as they could. In the Air Force, we called this "filling a midnight requisition."

One memorable impression was the high level of affection these old workers held for William Piper Sr., the founder of Piper Aircraft. They always referred to him as Mr. Piper and never said an unkind word about him. Two factors appear to be underneath this: First, the depression was in full swing when he made good jobs available to them; Second, he treated them very well and provided many facilities and benefits that were uncommon in the days of considerable labor-management strife.

John Vondran, from Stoughton, Wisconsin, had recently rebuilt his Piper Clipper. He was concerned about whether he had put the brake pedals in the right position on the cockpit floor. He cornered an old Piper employee and told him all about the brake pedal problem. The fellow listened patiently. At the end of the description, he told John that for 20 years all he did for Piper was weld left side landing gears. John was taken aback as he assumed that anyone who had ever worked for Piper must be an expert on everything associated with their airplanes.

That evening, there was a banquet, for about 200 persons, held in the pavilion. Bill and his father got tickets, but I was too late. I had to stand at the periphery of the tables to listen to the after dinner events. A grandson of William Piper gave a nice talk about the early days of the company. A long list of awards were given out for various categories of Piper airplanes. At the end of the awards, the three of us returned to the motel.

Around 10:00 A.M. Sunday morning, we rode in a van back to the airport. Bill, his dad, and I thought the Fly-in lasted through Sunday. Much to our surprise the airplane parking area was almost empty. On the way out to our planes, we encountered 75 year-old Clyde Smith, Sr, who was the test pilot for Piper during most of the production run of the Cubs. He is an icon to Cub owners. He can keep people entranced for hours with tales of his years at Piper. We talked to him for more than an hour. By the time we finished talking to Clyde Smith, it was afternoon. We hated to leave him, but we really needed to get on our way.

After taking off, a 1,300 foot climb was required to get enough altitude to clear the high ridge west of town. Once over the ridge, we flew across the twenty-two miles of solid forests again. They did not elicit quite as high a pucker factor as on Thursday. Bill and I overflew Clearfield again and went to Punxsutawany. Taxiing to the ramp, the field appeared deserted. Again, I had visions of having a fuel problem. We climbed out, started looking around. Bill found a fellow, in the large hangar, repairing a twin engine airplane, but he did not have a key to the fuel pump. In a few minutes, the woman, who had solved our fuel problem previously, appeared again and we were able to fill our Cubs' tanks.

Afterwards, the fellow from the big hangar gave Bill and me some small wood figures of Punxsutawany Phil, who was the focus of their local groundhog day celebration. Each February, Phil and a ground hog, named Fred, from Sun Prairie, Wisconsin, vie to see who correctly predicts whether the next six weeks will be winter or an early spring.

Bill and I retraced our flight path back to New Castle, but now the FBO had no fuel at all. Fortunately, both Bill and I had enough fuel to fly to the airport at Salem, Ohio. As we approached the area it appeared as though there were two airports. What looked like a north-south runway at the first airport turned out to be a drag strip. The FBO at Salem said that quite a few people had actually landed on the drag strip thinking it was the airport.

After topping off the Cubs at Salem, Bill and I had to fly around Canton again. By the time we arrived at Ashland, it was after 6:00 P.M.. The three of us decided to camp here overnight. Mr. Johnson let us use his car again and left the building open overnight. He was very accommodating given that, between the two Cubs, only a total of about 12 gallons of gas was purchased.

As Monday dawned, the yet-to-rise sun created a band of yellow haze at the eastern horizon. As I took off, the reflection of this band could be seen on the bottom of the Cub's right wing. High in the sky, there was a thin cirrus cloud layer that was highlighted by the hidden sun's rays. There were patches of light gray ground fog hanging over the fields and woods. Once the sun poked above the eastern horizon, the up-sun fog patches turned a beautiful golden color. The down-sun sides of the groves of trees remained almost black in color and were bordered by grey fog. It reminded me of a similar morning in Mankato many years ago, when I started out on a Cub ferry trip to Denver.

When we landed at Priebe, they were out of aviation gasoline. Bill and I had to fly the Cubs 12 miles south to the Findlay, Ohio airport. It was owned by the Marathon Oil Company, had two very long concrete runways, a brick operations building, and a brick FSS building. After refueling the Cubs, we went to the FSS to check weather. It didn't look too impressive to the west, but seemed to be decent Cub weather.

After takeoff, Bill and I flew a short northerly leg to get back onto the original Priebe to Warsaw line of flight. Now, the attention to details on the way to Lock Haven started paying off. By the time we approached Warsaw, the ceiling was getting low and it had started to drizzle. Yet, the Cubs arrived there right on our ETA. While the three of us were waiting out the weather; a fellow invited us to look at his Staggerwing Beechcraft. It was an immaculately restored C model done up in white with blue and gold trim. Seeing the Staggerwing made the several hours wait worthwhile.

Finally, we could see a smokestack that was three miles away and the weather was flyable. A four-lane highway and a parallel railroad track went northwest to Plymouth, Indiana and then to Valparaiso. Each of these towns had an airport. If the weather closed in, we could land quickly. I took off first with Bill and his father about a quarter mile behind me. I flew at about 100 feet above the highway, high enough to clear the many power lines, and low enough to keep well clear of the clouds.

By the time the Cubs passed Plymouth, the weather was deteriorating rapidly. Bill pulled up just behind me and 100 yards to the right. Five miles before Valpariso the visibility dropped to a half-mile, in heavy haze and drizzle, with a 400foot ceiling. Looking at the map, I saw that a north-south power line crossed the highway then went past the east end of the Valparaiso airport's runway. If I followed the power line, I could find the airport. In a few minutes, the powerline appeared out of the haze. I made a steep right-hand turn, and hoped that Bill could follow me. As soon as I rolled level over the powerline, I saw the lights on the east end of the runway. I was surprised as from the map I thought the airport was at least a mile further up the line. I threw a quick left turn, chopped the power, landed as short as I could, and went off onto the first taxiway. In back of me, Bill saw what I had done and landed further down the runway.

As I taxied in, I recognized three Cubs, a Stinson 108, and a Citabria belonging to the contingent from the Des Moines, Iowa area. The leader of the group was a tall fellow from Blakesburg, Iowa, who flew a rather beat-up Cub. He was not very talkative and was quite hard to draw into a conversation. The group had left Lock Haven early Saturday morning and stopped here when the weather went bad. They were surprised that Bill and I made it this far. About five minutes after we landed, a Piper Cherokee made an IFR approach. Then, a DeHaviland Chipmunk landed. The Chipmunk pilot and his wife had been scud running, from the west, on their way to Montreal, Canada. When the pilot saw the runway lights, he landed immediately. If either plane had been a few minutes earlier, it might have been very "interesting" for all of us.

The next morning (Tuesday) the weather was not a bit better. The three of us took a cab back to the airport with some of the Iowa boys who also had opted not to camp. By 11:00 A.M., the clouds started to lift and break up, so we all climbed in our aircraft. It was an impressive sight to see five Cubs and the two other "taildraggers" taxi out swinging their noses side to side to see ahead. It reminded me of movies of B-17's lumbering along the taxiways at the beginning of a WW II bomber mission. Once in the air, we formed a loose gaggle of airplanes.

The 17-year-old fellow, flying the Citabria, was only a student pilot. His flight instructor had signed him off for the cross-country trip to Lock Haven with the group. Because of this, they made him lead the flock. When he got off the course, the other four dutifully followed him. Few student pilots have had as a memorable first solo cross-country as he did. In twenty minutes, the Iowa planes were well ahead of us on our way to Morris, Illinois rather than Joliet. Morris was on the direct route to Des Moines and has an on-field restaurant. By the time we were past Joliet, the clouds disappeared and it was a beautiful day. In the distance, I could see the Iowa planes landing at Morris. Bill and I arrived five minutes later.

At the gas pit, I realized they had been pushing their engines hard. I took four gallons while their Cubs used six gallons each. They had a case of get-home-itis. The airport mechanic was not very happy about gassing up seven airplanes with only 4-6 gallons apiece. The whole time, he grumbled about being a mechanic not a lineboy.

After the group had lunch at the airport restaurant. The Iowa planes departed to the west. The three of us climbed into our two Cubs, took off, and headed north for Harvard. After an hour and fifteen minutes, Bill and I landed at Dacy's and refueled the Cubs. From here it was only 72 miles to Waunakee. The Cubs practically flew themselves home. The 1,446 mile round trip took 25 hours and 22 minutes of flying time at an average ground speed of 57 mph.

After we landed, the three of us talked for a few minutes. Bill's father told Bill:

"When the weather gets really bad, follow Frank."

A few minutes later they took off and I watched their Cub slowly disappear into the distance. It was an enjoyable trip, hopefully, we could do it again.

Chapter 14
To Lock Haven Alone

In the spring of 1991, Bill Coleman told me that he would like to go to Lock Haven again. As the departure time drew near, his going on the trip became questionable. Finally, Bill told me that he could not go. I would have to fly there alone for the first time.

The maps were studied again and I made a few changes to the "80-octane" route. In order to shortened the leg to Warsaw, Indiana, the Sanger, Illinois airport was to be used instead of Joliet. Because of the problems associated with getting fuel at Punxsutawany, I planned to fly directly from New Castle to Clearfield, Pennsylvania that now had a fixed base operation. These changes increased the distance to Lock Haven by only six miles.

The Cub and I departed on Wednesday, 14 August at 7:30 A.M.. This time the weather cooperated with clear skies and light winds. The Cub was refueled at Dacy's just as they were starting the business day. After leaving Dacy's, I followed a road south to Marengo and then to Plano where the course line turns to the southeast towards Sanger. Just past the turn point, my peripheral vision alerted me to something underneath the Cub. I looked down and saw a yellow Sikorski heavy lift helicopter climbing up directly underneath me. I made a tight turn to the right and moved out of its way. This type of helicopter is often used to lift air conditioning units onto the tops of buildings under construction. Later, I learned that, on their cross-country flights, these choppers carry 55 gallon drums of fuel in their cargo bay. Apparently, this chopper had landed in the farm field to refuel from the drums.

The Sanger airport is somewhat difficult to find as it is not located near any city. In addition, the main runway is somewhat hidden as it goes downhill to the east. There are several rows of aged corrugated metal hangars with faded orange roofs, a large old maintenance hangar, and the airport office is in what once was a farm house. The two story white house has a small office just inside the front door. The east-west runway is a narrow strip of crumbling asphalt with low weed covered embankments on each side. There is also a northeast-southwest runway in about the same condition. All-in-all not the nicest airport on the trip, but they had 80-octane fuel.

The gasoline pumps were located on a small ramp near the farm house. At the pumps, the Cub sat with its tail quite a bit lower than when on level ground. After refueling the Cub, I pulled the prop through a few blades, turned on the switch, and gave the prop a quick flip. Then a sound like a shotgun being fired came from underneath the airplane. I had visions of the Cub going up in flames, but a quick inspection revealed that it was still in good shape. I could not see anything, which might have caused the explosion, gave the prop another flip, and the engine started normally—another one of life's little mysteries.

From Sanger to Warsaw, Indiana, the new route initially was slightly south of the previous one. Twelve miles east of Sanger I flew low over a large lake with housing completely around it. On its southeast side there was a 1,000foot tall television tower that I looked up at. Some 40 miles further, I flew right over the Starke County airport, but again no activity could be seen. The Cub crossed the Valpariso to Warsaw highway and railroad track at the village of Bourbon that has a well marked water tower. It was only 14 miles from there to the Warsaw airport.

Several major companies are located in Warsaw and their corporate jets are housed in two large modern hangars. To accommodate the jet traffic, both the north-south and east-west runways are very long. The Warsaw airport has a nice operations building with a well appointed pilots lounge.

After lunch, I strapped on the Cub again, taxied to the south, and took off on runway nine. As I broke ground I heard a Lear Jet call in on a five-mile final to runway 27--we were face to face. I quickly made a turn to the right and rolled out about 100 yards south of the runway. By the time the Cub's wings were level, a second call was heard on the radio announcing that the Lear Jet had landed and was turning onto the taxiway. I did not see the Lear Jet go past me. Strangely, the pilot of the Lear Jet made no mention of the Cub being in front of him. I wondered if he even saw me.

The first 25 miles east of Warsaw contains a number of small lakes. The one nearest the city has houses and swimming beaches on its shores. The cluster of lakes further on are less developed. These latter lakes are quite narrow and their shores are mainly wooded. Just south of a pair of these "finger" lakes was a white VOR building that enabled me to get the Cub well established on the course line. Then, I flew past the wonderfully named town of Churubsco. It took another hour to finally reach the Priebe airport. This time I was going to land to the southwest. Final approach was over the landfill and the Cub touched down where the runway went through the trees. The runway still looked too small for a P-51 to use.

The 76 mile flight from the Priebe airport to Ashland, Ohio, on a heading of 105 degrees, was a routine one. I landed 6:35 P.M. and set up my tent to camp overnight here. The airport car was used to drive the several miles downhill to the edge of Ashland. I went to the same restaurant that Bill, his father, and I had dined at on the previous trip.. Once I was settled in for the night, airplanes kept landing until around 1:00 A.M.. Until then, I didn't get a lot of sleep. Finally, things calmed down and I slept until 6:30 A.M..

In the morning, I was able go inside the warm airport office to eat my breakfast items. Around 8:00 A.M., I packed up the Cub, wiped the dew off the windshield, propped the engine, took off, and turned into the rising sun. The Cub and I flew across the large valley at Wooster and then over the high school again. Just west of Massillon, Ohio is the village of Dalton that has a tall, blue cylindrical water tower perched on top of a small hill. Although there are numerous checkpoints in the vicinity, the water tower is a welcome sight.

On the south edge of Canton there are several industrial complexes that appear to be steel fabrication facilities. Over the VOR building on the hill south of Canton, a new heading of 75 degrees, a course heading of 68 degrees plus seven degrees of west magnetic variation, was followed. The route of flight took the Cub over the city of Salem and then just south of the Youngstown-Elser airport. I circled around the New Castle airport and set up for a landing on runway 23 that goes slightly uphill. A downdraft was expected where the northeast end of the runway ends at deep valley, but none occurred.

After refueling the Cub at New Castle, I flew east toward Clearfield–exactly 104 miles away. Clearfield was an attractive stop as it now had fuel and was only 52 miles from Lock Haven. If I ran into weather at Clearfield Mountain, there are alternative airports at Du Bois and Punxsutawany some 26 miles back to the west. The former was 10 miles north of this new course line. The latter was only 5 miles south of the course line. Punxsutawney was unattended and presented a fuel problem, but "any port in a storm."

Scud-running on their way to one of the early Lock Haven fly-ins, a group of Cubs landed at Du Bois. The newly-minted FSS staffer there promptly started to report them to the FAA for flying in non-VFR conditions. When the veteran station manager got wind of this, he told the staffer "those are Cubs, they always fly in that kind of weather." He quickly canceled the report. Today, my good luck with the weather was holding, there were clear skies, tailwind, and comfortable temperatures. Hence, an alternative airport was not needed, but it was comforting to know that they were available.

The present route was north of the route that went to Punxsutawany. Consequently, it possessed a whole new set of checkpoints. Thirty-six miles east of New Castle, the Cub flew across the Allegheny river at the hamlet of West Monterey. That made me wonder where East Monterey was. I couldn't find it on my map. The tall smoke stack at the power station, a few miles to the south, could be seen again sticking well up above the river bluffs. I flew past a series of hamlets-- Rimersberg, Cool Spring, and Syksville. Beginning at Cool Spring, the countryside has many well defined hills with tightly rounded tops with narrow valleys between the hills. Five miles past Syksville, there was a hamlet, unnamed on the map, perched on top of the north end of a ridge. Above it was a 354foot tall radio tower. Two and one half miles to the east of the hamlet, there was a railroad at the bottom of a deep valley that cut through the southern third of Clearfield Mountain.

At this point, I started climbing the Cub to gain altitude. In a few minutes it was at 2800feet that is high enough to easily clear the top of the mountain with its radar domes and radio/television towers. The top of the mountain is heavily wooded and is a Pennsylvania state park. Several additional towers are located about halfway down the east side of the mountain. I had to stay clear of them as I let down to the airport that was 894feet lower than the top of the mountain.

Landing at Clearfield is interesting as the runway is on top of a ridge overlooking the town. It looks much like a mesa-top airport in the southwest U.S.. Interstate 80 runs alongside the airport, but is some 300feet lower than the runway. I was pleased to discover that there was a small restaurant in the airport's main building. I had a cheeseburger and a glass of milk for lunch, a rare delight for a cross-country Cub pilot.

I elected to take the straight across route to Lock Haven again. This time the first 30 miles looked much less formidable than in the past, however, the last 22 were still quite spooky. Five miles past the hamlet of Orviston, I could see a high east-west ridge several miles to the south with a large V shaped cut through it. There was a creek at the bottom of this cut that ran to the village of Beech Creek that is in the same valley as Lock Haven. Despite this escape route, I stayed on the direct line of flight to Lock Haven.

In less than an hour, the Cub passed over the last high ridge. Then, I started my let down into the valley. I landed, to the east, on the grass runway. A fellow on a motor scooter directed me to one of the far west parking rows. Not a very desirable location, but it would have to do. I had flown from Madison to Lock Haven in 10 hours and 22 minutes of flying time at an average ground speed of 68 mph, which is essentially a no wind situation.

After the usual drill of getting registered, I walked around the flight line. The theme of the Fly-In this year was "The Piper Cub goes to War." There were a half dozen beautifully restored L-4's parked near the entrance. A re-enactment group from Pittsburgh had set up a typical WW II Army tent and command post. All the personnel wore authentic WW II uniforms. The jeeps and trucks were just as they were in the war. The over all effect was very impressive. Across the field, a B-17, named "Sentimental Journey", was parked on the ramp. It had been at the National Warbird Museum air show at Genesco, New York. Because of the name of their B-17, they decided to stop at Lock Haven on the way back to their home base in Texas. Later in the day they took off and made several very nice low passes down the runway.

Late in the afternoon, one of the vans took me to the Comfort Inn (formerly the Holiday Inn). On this trip, the driver was a retired Piper employee who told me an interesting story. While he was still working at Piper, he had lost his medical certificate due to a heart condition. An FAA man from Washington, D.C. was at the Piper plant to run some tests on a new plane that Piper was getting certified. Since, the FAA man was not checked out on the plane he asked if someone would fly him while he ran the tests. The driver volunteered, so they went up and flew the tests. That night his wife asked him what he did at work that day. He replied: "flying tests for the FAA." His wife knew that he was not legal and was horrified. The driver though was smiling like the Cheshire Cat and the FAA was none the wiser.

Friday morning, much to my disgust, I discovered that the Cub's battery was dead. After parking the Cub, I had forgotten to turn off the master and battery switches. This was despite the "Switches Off" placard above the outside of the left window that reminded me to turn off the switches. I had put the placard there after my daughter Ann had left the switches on after a flight. Obviously, I did not look at the placard or its message did not register when I closed the left window. Digging out my tools, I set forth on the task of removing the battery and then charging it overnight in the FBO's workshop.

Saturday afternoon I was sitting by my Cub when a couple wandered past. I asked them if they knew the Dempsters from Vashon Island, Washington whose Cub was parked a few rows from mine. I was surprised to learn that I was talking to the Dempsters themselves. Bob is a sculptor and with his full beard looks the part. Diane works for Boeing in some capacity, when she is not out "Cubbing" around the world. From my English Fly Past magazine, I knew they had taken part in the London, England to Melbourne, Australia air race. They had shipped their Cub to London, reassembled it and entered the race. Their Cub has an 85hp engine. With wing tanks their Cub has a total of 36 gallons of fuel onboard. This enables them to stay aloft six to seven hours.

They flew from London to Germany to Yugoslavia and then to Greece. After a bit of island hopping they flew over water for five hours to Egypt. Bob related flying over Egypt at 4,000feet when, due to the density altitude, the Cub literally fell out of the sky. He happened to be right over an Egyptian Air Force base at the time and was able to make a good landing. The Egyptians told him it was 125 degrees on the ground. His chances of getting the heavily laden Cub airborne again at the local density altitude were nil. He and Diane decided to call it a trip. They took the wings off the Cub, put it in a container, and shipped the Cub to Miami by sea.

I told them about my and Ann's narrow escape in Montana. They were also told about the landing on the sod farm when the engine broke a rocker arm. They seemed to be more impressed with the Montana story than with their own adventures in Egypt. In the spring of 1993 they bought a new Piper PA-18 Super Cub at the factory and flew it home to Vashon Island, Washington. Once home, they started to outfit it for a flight around the world. They have been sending out a periodic newsletter describing their efforts. Bob has been working on his instrument rating in parallel with outfitting the plane. They sent me a yellow decal with the information about the trip, which I put in the Cub just to remind me that there are still some real adventurers left in this world.

One of the Fly-in's forums was about flying L-4's in WW II. It was very interesting to listen to pilots who have received very little recognition for performing some very dangerous flying. I talked to Mike Strok who flew L-4's in Italy during WW II. He had been in Korea in a non-flying capacity. He was familiar with much of the Korean front line that I had flown over. I told him about the time when my pilot spotted an L-19 below us. We peeled off into a high speed dive and at a low altitude our B-26 went whistling past the L-19. My pilot pulled up and made a wide turn to get lined up behind the L-19 again. About this time, the L-19 pilot started flying north through a series of valleys with us catching up with him from the rear. I was so engaged in watching the L-19 zig and zag through the hills and valleys that I did not pay a lot of attention to where we were being led. Suddenly, the L-19 made a quick 180 degree diving turn. As we zoomed by, I realized we had flown into North Korea at a very low altitude. The Chinese must have been caught by surprise as nobody started shooting at us. My pilot made a tight turn to get back into South Korea. I'll bet that the L-19 pilot had a good laugh when he told his pals what he had done to us.

At the Banquet Saturday night, I happened to sit across from two young women one of whom was a

Continental Airline pilot. The other was an air show pilot. The two of them also had flown in the London to Melbourne race. They had flown as far as Jordan where the engine on their Piper Tri-Pacer failed. After a lot of bureaucratic red tape and the intervention of King Hussein, they were able to ship in a new engine. Long after the race was over, these two gals continued to have a lifetime of adventures along the route. They would get a few weeks of vacation from their jobs. Then, fly the Tri-Pacer as far as time allowed, put it in a hangar, and go back home. This process was repeated a number of times. Finally, after eighteen months of this they reached Australia, sold the Tri-Pacer, and went home for good.

Sunday morning started out with low ceilings and some clouds down in the valley. As I rode in the van to the airport, I saw a Piper J-5 following Interstate 80 to the west. I decided that if he could make it so could I. In the past, I had talked to a number of Cub pilots who followed the Interstate west to Clearfield. Even Clyde Smith Sr said this was the preferred route to the west. The ceiling was just above the tops of the ridges. The visibility was about 3 miles in heavy haze, so the Interstate route seemed like a good idea.

The first leg is up a broad valley to a small town where the highway makes a 90 degree turn to the right and starts to climb up the face of a tall east-west ridge. Shortly, the highway was in a rather narrow slot going up the ridge. The clouds were just barely above the top of the ridge. The Cub was getting hemmed in to the point where it looked as if a precautionary 180 degree turn would be a difficult maneuver. To compound matters, the highway kept climbing up toward the cloud deck. Every time I climbed up a bit to see where I was going, the Cub would get into the clouds. Then I would have to drop down 50feet to get clear of the clouds. Just about the time I decided to get out of the muck and head back to Lock Haven, the town of Snowshoe came into view. The Cub had reached the summit of the ridge and I could see the Interstate highway heading downhill. The Cub and I had made it over the ridge. I was glad that I did not have to fly back toward Lock Haven into the oncoming traffic of other people following the highway.

Just past the summit there was a power line across the highway. At last fall's Grassroots Fly-In I talked to several persons who had been to Sentimental Journey in 1990. They said that the weather had gone bad several days before the Fly-in started and stayed that way the whole time. The Iowa contingent had left their airplanes in eastern Ohio and had taken a Greyhound bus to Lock Haven. There was only one Cub pilot who made it in from the west. The story was that he was following the Interstate highway from Clearfield to Lock Haven under a very low ceiling. As he reached the highest point of the highway, at the village of Snowshoe, he saw a power line across the road. The lines were too low to fly under so he pulled up into the clouds, counted to three and let back down again. That fellow had more guts and fewer brains than the usual Cub pilot.

From Snowshoe to Clearfield, the visibility was three miles in haze and the ceiling was slowly increasing. I was flying about 100feet above Interstate 80 and following it downhill to the west. The Clearfield airport was almost missed as the Cub and I passed well below it. Fortunately, I happened to be looking up and saw the roof of the main hangar as I flew past. A bit of power brought the Cub up above the runway and I landed back toward the east. I taxied up to the gas pit, killed the engine, and started to look for someone to gas up the Cub. It was 9:30 A.M. and there were no signs of life anywhere. Some thirty minutes later, one of the local pilots showed up, but he did not have a key to the gas pumps. He and I searched through the desk drawers in the airport office and in the shop, but could not find a key. Finally we went back to the pumps and discovered that the 100LL pump was not locked. I gassed up the Cub, went back into the building, checked on the price of gas, and left enough money to cover the fuel.

The local fellow said that it was a good idea to get out of here as soon as possible as the airport would quickly get socked in for the rest of the day. A few weeks after getting home, I talked to Don Johnson. He and

Steve had seen me depart Lock Haven and decided to follow. It took them about an hour to pack up the tent and get off the ground. They made it to Clearfield about 10:30 A.M.. As soon as they landed, the conditions went to zero-zero just as the local pilot had predicted. They spent the night in a motel and departed midmorning the next day.

To the west, I could see the radar domes on top of Clearfield Mountain as well as the radio towers on its east side. I took off and started to follow the road that goes up to the radar domes. At about 1,900feet, I was in the clouds and made a 180-degree turn back into the valley. I found Interstate 80 and started to follow it up the ridge. Again at 1,900feet, I was in the clouds and had to abandon the attempt. I knew that, when Bill Coleman and I came here the first time, we had flown around the south end of the ridge. I decided to mosey down that way to see if an end run was possible.

Nine miles to the southwest of Clearfield, there is a valley going northwest into the mountain. Checking the map, I saw that the valley went most of the way through Clearfield Mountain. At the bottom of the valley there is a railroad track to follow. I turned into the valley whose sides were completely wooded with unusually dark green trees. The cloud deck was just brushing the trees on the tops of both sides of the valley. Wisps of clouds were sweeping through the trees on the sides of the valley. I flew up the valley for seven miles. Suddenly, I saw a radio tower a mile to my left. It was at the unnamed hamlet that I had flown past on the way to Lock Haven. The tower was on the west side of Clearfield Mountain. I made a left turn at the tower that put the Cub right back on the course line to New Castle.

In ten minutes, I was in clear skies with enough fuel to by-pass New Castle and fly to the Youngstown-Elser airport. Their single east-west runway is quite long, but the west end is roughly 60feet lower than the east end. I had to land downhill and touched down as close to the east end of the runway as I could. I was almost to the west end when the Cub finally stopped rolling. I back-taxied the runway and went to the gas pit. After topping off and paying the fuel bill, I went into their shop. It was filled with antique airplanes. There were a couple of Beech Staggerwings, a Stearman, and the serial number one Travel Air Mystery ship that is one of the famous racing planes of the 1930's. The A&I was a trustee of the Staggerwing museum in Tullahoma, Tennessee. He was restoring the Mystery Ship for the museum.

The next leg was to Ashland, Ohio. When I landed there, the Piper J-5 that I had seen taking off at Lock Haven was parked on the ramp. Its pilot was a fellow, from Oklahoma, whom I had conversed with at Lock Haven. A few years later, I met him again at a Fly-in at Bartlesville, Oklahoma.

The next stop after Ashland was Putnam County Airport a few miles west of Findlay, Ohio. As I flew over Findlay, I debated about landing there, but did not. They only had 100LL while Putman County was listed as having 80-Octane. Putnam County had a long east west runway and a brand-new operations building. After landing, I taxied up to the gas pumps. The airport office building was open, but deserted. After some 45 minutes, a family in a Piper Cherokee landed and taxied to their hangar. I went over to there and told them my predicament. Fortunately, the father had a key to the pumps. He sold me enough gas to fill the Cub's tank.

They departed and the fun began. I flipped the prop, but nothing happened. It was flipped again and again and still no life. After about an hour, I decided that the magnetos must be hot like Bill Vogelsang's were on the Gastonia trip. I pushed the Cub over to the building and ran the water hose on the engine for quite a long time. Then, I pushed the Cub back to the ramp. The prop was flipped some more, nothing, not even a pop. Finally, a local fellow, in a pickup truck, drove in to the airport. He flipped the prop for another half hour to no avail. In the mean time, the sun was rapidly descending in the western sky. Another local showed up who owned a Cub and a Luscombe. He had a wrench big enough to go around the bowl screen on the Carburetor. I used it to loosen the plug enough for gasoline to run out at a good flow. After draining about a quart of gas, I tightened up the screen plug and safety wired it in place. I later learned there is a small plug on the bottom of the carburetor just for draining it.

I flipped the prop and got a few feeble pops out of the engine. The Luscombe owner theorized that the low speed carburetor jet was plugged. He thought that starting at a higher throttle setting might work. He climbed in the back seat, set the throttle at nearly full power and stood on the brakes. I flipped the prop from its rear and the engine started with a roar! It had been three and one half hours since I had landed. All hope of getting home in one day had vanished, however, Plymouth, Indiana looked possible. I thanked the two helpers and took off into the setting sun. An hour and 40 minutes later, I landed at Plymouth just as the last light of the day disappeared. On the ramp there were an L-4 and a J-3 that had been to Lock Haven. I tied the Cub down in back of them. Then, I followed the asphalt path from the ramp to the Holiday Inn that is adjacent to the west end of the runway.

The next morning dawned clear and calm. After checking out of the motel, all three of us Cub pilots ended up on the ramp waiting for the airport office to open. Around 8:30 A.M. the FBO finally drifted in and we refueled our airplanes. I took off first, followed by the L-4, and the other J-3 was last. For about ten minutes, we flew in a loose gaggle. They slowly pulled ahead of me and by the time the Joliet, Illinois airport was in view they were nowhere to be seen. I refueled at Joliet. After another hour and 40 minutes of flying time, I was at Dacy's. It was mid-afternoon when I landed at Waunakee.

The trip back took 11 hours and ten minutes of flying time at an average ground speed of 64mph. I had light winds nearly the whole time. In addition, once west of Clearfield Mountain, the entire return trip was in clear blue skies--a rare treat east of the Mississippi River. Once home, I removed the carburetor and disassembled it. The fuel bowl had about a half-teaspoon of fine grit in it. Somewhere along the trip, some contaminated fuel had been put in the Cub's tank. I cleaned out the grit, reassembled the carburetor, and put it back on the engine. After that, the engine started normally.

West Branch of the Susquehanna River

Cub over Reclaimed Coal Mine in Pennsylvania

Lock Haven, Pennsylvania Airport

Fog in Valley of the West Branch of the Susquehanna River

Chapter 15
Lock Haven Again

In 1992, I was not planning to go to Lock Haven. However, my daughter, Ann convinced me that I should fly the Cub to Lock Haven while she drove there from Wilmington, Delaware. Once committed to the trip, I talked to Bernie Neuman who has been my friend since high school days in Mankato. We also were roommates, for a year, at the University of Minnesota. Bernie had retired from the Honeywell Corporation were he was an aerospace engineer. I told him about the Lock Haven trip and invited him to join me. He readily agreed to go along.

This trip was to be the first cross-country with my new radio. The original NARCO radio had worked perfectly on the Prescott trip, but its performance had been steadily degrading over the years. It had reached the point where it was more of a hindrance than a help. Finally, I purchased a TKM 11 radio that simply slid into the existing radio tray as it was plug to plug compatible with the NARCO. It was a pleasure to be able to both hear and transmit clearly.

At 8:00 A.M. on a gloomy Wednesday morning (August 11, 1992), Bernie and I loaded up the Cub. A line of thunderstorms could be seen a few miles to the northeast of the Waunakee airport. Off in the distance, the storm clouds were pierced, on a regular basis, by bright flashes of lightning. Since the thunderstorms were not on our line of flight, we took off with Bernie in the back seat as navigator. Once past the city of Madison, I dropped the Cub down to 200feet AGL and then our troubles began. Suddenly, it became almost as dark as nighttime, the lightning flashes were still off in the distance, and the visibility was good. However, the clouds above us had turned black, the wind became very strong, and severe turbulence was encountered. The Cub was being tossed up and down as well as side to side. A wing would go up, I would slam the stick to that side in an attempt to get the wings level. Then, the Cub's nose would swing to one side. The opposite rudder pedal would be pushed to the firewall to contain the swing. While this was happening, the nose of the Cub would rise up. I would push the stick forward and the nose would drop. It seemed as if I was stirring a pot with the control stick and doing a clumsy dance on the rudder pedals. The net effect of all this activity was that Bernie and I felt as if we were riding inside a concrete mixer. After being pounded for about ten minutes, the turbulence began to subside and the Cub was starting to fly with some semblance of normality.

A mile or so past the village of Stoughton , the weather finally calmed down and Bernie said that we had acquired a better ground speed. The Cub plodded along to the Dacy airport at Harvard, Illinois. After refueling there, we took off to the south into some not too promising looking weather. There were several layers of clouds that combined with the heavy haze put the Cub in a grey cocoon. In addition, it was quite cool for this time of year.

A few miles south of Dacy's, Bernie and I encountered a broken scud layer whose bottom was at about 300feet and its top was at about 500feet. Since the clouds were broken, I climbed up on top through a hole in the clouds. Above this lower layer, there was a second dark grey ceiling at about 2,000feet stretching from horizon to horizon. It quickly became apparent that to the south of our present position the top of the lower cloud layer was building higher. I had no intention of climbing over this cloud ridge only to find ourselves above a solid overcast. I found a another hole in the scud, made a diving turn down through the clouds, and leveled out at 100feet off the ground. Down at this lower level the visibility was reasonable, but Bernie and I had some trouble seeing checkpoints that were not directly on the course line. We never did see a radio tower that was only a half-mile away.

As the Cub flew over Plano, the lower scud layer ended and the visibility increased to 5 miles. We were able to pick up the course line to Sanger, Illinois. Approaching the Joliet airport, Bernie turned on the radio and we could hear a number of pilots talking. A couple of them were in the vicinity of the Joliet airport. With a lot of visual scanning of the airspace, Bernie and I were able to spot the nearby airplanes. A major problem with listening to the common airport radio frequencies is that you can hear all the transmissions within a fifty-mile radius. We had to listen carefully to determine which ones were relevant.

As I refueled the Cub at Sanger, Bernie noticed gasoline running out of the exhaust shroud. He reached into the cockpit and quickly shut off the main fuel valve. After I paid the gas bill, Bernie climbed in the front seat and I propped the engine. On the first pull there was a sound like a shotgun being fired. I jumped back and looked around for a fire, fortunately there was none. After some discussion, Bernie and I concluded that gasoline vapors in the exhaust shroud had exploded. The nose high position of the Cub on the ramp at Sanger resulted in the carburetor float valve sitting in a partially open position. This allowed gasoline to flow into the venturi, down into the airbox and then into the exhaust shroud, where it was ignited when the engine started.

Since we were past the bad weather, Bernie was to be the pilot on the leg to Warsaw, Indiana. Although he had flown with his son many times, Bernie had no previous stick time. Once in the air, I showed him how the controls worked and after a few minutes he took over. At first, the plane would buck and weave as he over controlled. When the excursions became extreme, I would put things back to normal. After a half hour of this, he was keeping the plane on a reasonably even keel and headed in the right direction.

Bernie noticed that as we burned fuel the Cub would get out of trim. With a full tank the Cub's center of gravity(CG) was at its nominal position. As fuel was used, the weight of the spare quarts of oil, tools in the baggage compartment, and our luggage caused the CG to move aft. As a result, the elevators become quite sensitive. Bernie grumbled about this. He wanted to move some of the weight forward, but there was no place to put it. He became resigned to using the trim tab and keeping some forward pressure on the stick.

In addition to the CG shift, the Cub's primitive fuel gauge bothered Bernie. As fuel was used, the wire attached to the cork float descended in a nonlinear manner. For the first third of the tank it fell quickly. In the middle third, it fell slowly, and in the last third it fell quickly again. At Honeywell, Bernie had designed highly accurate fuel gauges for large airliners. The Cub's nonlinear system offended his sense of engineering. Worst of all, it didn't even have a gauge face with numbers on it.

Flying from the back seat, I landed the Cub to the south on Warsaw's long north-south concrete runway and taxied to the ramp. Bernie and I took the courtesy car to Grandpa's restaurant that is only a mile or so from the airport. The interior of this restaurant is mostly well worn varnished dark pine and is poorly lighted. It has all the ambiance of a truck stop, but sits alone alongside the highway. The food was a notch above the ubiquitous Stewarts's microwave sandwiches.

After lunch, Bernie switched to the Cub's back seat as I wanted to be in front when landing at the Priebe airport in Ohio. To give Bernie a better view of the countryside, I flew the Cub rather high, 800feet AGL. At this altitude there was a nice tailwind to help us along. In the midst of the lake country east of Warsaw, the Cub flew over the VOR building and we were right on course. In another 23 minutes we went past the rectangular lake at Grabill and then the Maumee river at Antwerp, Indiana.

The thousand-foot tall television tower west of the Priebe airport could be seen from many miles away and heading the Cub toward it led us to the airport. This time I landed to the northeast, over the telephone wires, and rolled out in the slot through the trees. The runway still looked too narrow to land a P-51. Due to EPA regulations, the airport's underground gasoline tank had been replaced by an above ground tank. The local P-51 was parked in front of it. Bernie and I rolled the Cub nose to nose with the P-51 and topped off the Cub's tank with 7.2 gallons. In the shops, they were restoring a Republic P-47. The F4U Corsair that was here three years earlier had been completed and flown out.

Bernie went back into the front seat. After takeoff, he took over the piloting chores. By this time, he was getting reasonably smooth on the controls, so I only had to sneak in an occasional correction. On this leg, there were a number of power lines running from the northwest to the southeast that, when paired with a railroad track or a highway, provided excellent checkpoints. Twenty-two miles from Ashland, Ohio we flew over Plymouth, Ohio and, to the south, the tall buildings in Mansfield, Ohio were visible. A few miles further, the broad eastern slope of the valley in which city of Ashland is located could be seen.

Bernie and I reached my favorite stop, Ashland, at 6:00 P.M., refueled the Cub, and proceeded to start the engine. Much to our disgust, it would not start. About two hours of daylight remained and we could have reached the Youngstown-Elser airport or even New Castle, Pennsylvania. The longer we propped the engine, the fewer signs of life it showed. Finally, it became clear that we would not leave Ashland tonight, so we gave up. Mr. Johnson, gave us a ride to a local motel and offered to pick us up in the morning. I am continually amazed at his hospitality as he certainly did not cover his costs on the meager amount of gasoline the Cub took.

The next morning, Mr. Johnson showed up right on time and we went back to the airport. During the conversation, I learned that he had been a flight engineer on B-29's flying out of Clark AFB in the Philippines. For Christmas this year, I sent him a 1945 book about the development of the B-29.

Since it was cool this morning, the Cub started on the first flip. Delighted, Bernie and I climbed into the Cub and by 8:30 A.M. were headed for the Youngstown-Elser airport. Our line of flight took us across the four lane highway just east of the Ashland airport. In eight more minutes the Cub reached the valley at Wooster. This north-south valley is perhaps a mile wide and quite deep. There is a railroad track alongside a southbound creek and a few farm fields at the bottom of the valley. There also were several large man-made rectangular ponds in the valley that appeared to be recreational areas. As usual, we flew over the High School on top of the east side of the valley.

With two of us in the Cub one would expect a fair amount of conversation, typically, that was not the case. The usual interchange goes as follows:

"Bernie, I need a ground speed and an ETA to Elser."

"OK, I just got a time hack on the four lane highway. I will let you know when we reach the railroad track."

After ten minutes or so, Bernie comes back on the intercom with

"I got a ground speed of 63 mph. We should reach Elser at 9:11."

"I like the ground speed. Let's hope it holds."

These interchanges go on periodically all day long and about the only other conversation goes like this:

" Bernie, dig out the Fig Newtons, so I don't starve before we get to New Castle."

"How many do you want?"

"I'll take three."

"Here you are."

"Thanks."

Fig Newtons are a staple of life on cross-country flights in the Cub. Often the crew does not get to eat lunch at a decent hour. The Fig Newton's will subdue the hunger pains for a half hour or so.

Navigating around Canton, Ohio is very peculiar. You can see the city, there are number of small towns, numerous railroad tracks, and a large number of power lines. There is so much to be seen on the ground that it becomes indistinct clutter. Because of this, it is difficult to set up a heading and stay on course. The main problem is that little on the map seems to match what you see on the ground. It also is easy to confuse one checkpoint with another. The only clear-cut checkpoint is the city of Salem. Fortunately, the Cub flew right over the center of it. Just to the southwest of town is a grass strip airport tucked down in a valley. The main Salem airport is three miles north of town, but we did not land there.

Flying the compass headings Bernie gave me, we hit our Youngstown-Elser airport ETA right on the nose. Since the Cub's brakes are not very effective, I decided to land uphill to the east and also slightly downwind. The workshop was still filled with antique aircraft: four Beech Staggerwings, a different Stearman, the Travel Air Mystery Ship, and an Aeronca Champ. The Staggerwings are a cabinet makers delight with lots of beautiful wood formers and stringers. The workmanship on all the planes was first class.

Taking off to the west proved to be very "interesting." The Cub broke ground easily, but our rate-of-climb was not going to get the Cub over the power lines on the low ridge west of the airport. A shallow turn was made to the north (I should have turned south into the wind) to parallel the ridge. By keeping a small amount of back pressure on the stick, the Cub was made to slowly climb. After a while I was able to turn the Cub into the south wind and climb parallel to the power line. In a few minutes we had enough altitude to get back on course and head east toward New Castle, Pennsylvania.

Since gasoline availability had been rather iffy there in the past, New Castle was overflown. Bernie and I were going fly the Cub directly to Clearfield--117 miles away. In poor weather this could be a marginal proposition. Today's weather was excellent, so we did not anticipate encountering any problems. A few miles past New Castle and to the south of course there is a very high, east-west, wooded ridge. Interstate 79 goes south through a gap in this ridge. Because of the gap, we could see part of the very large Lake Arthur reservoir. To the north of the ridge, where we were flying, the land vaguely formed a very wide basin. The Cub was flying at about 500feet(AGL) where Bernie and I had a panoramic view of the countryside.

I tried to keep it a secret, but Bernie soon discovered that we were flying only a mile or so south of the famous Slippery Rock State Teachers College. For some reason, national radio and TV sports announcers love to report the scores of the school's football and basketball games. Seventeen miles further, we flew over a north-south valley. The little town of Petrolia and its oil refinery clung to the west side of this valley. There were numerous oil storage tanks, painted white, on several levels of the hillside. At the bottom of the valley there was a railroad track with a string of tank cars parked on it. The oil refinery seemed incongruous as we had not seen any oil wells in the area.

The Cub crossed the south flowing Allegheny River just north of East Brady. Eight miles further to the south, the power plant's tall smoke stack could be seen poking high above the top of the valley walls. To the east of the river, the countryside becomes quite hilly with many ridges and valleys. It is reminiscent of the coulee country near LaCrosse, but here the valleys are much deeper and the ridges much higher than in Wisconsin. The sides of most ridges are heavily wooded. There are small farms on the tops of the ridges and also down in the valleys

Fifteen miles east of the river is the town of New Bethlehem that is located in a valley. Despite the 100 yard rule (it used to be the 100 foot rule), the Cub was a bit off course when we reach New Bethlehem. A few miles past this town, Bernie and I flew over the railroad tracks that had so intrigued Bill Coleman. They went in and out of tunnels and over long iron railroad trestles. The latter appeared to be of the same design as those in Montana and Nebraska. A little while later, Bernie and I flew past the radio tower at the northwest end of the valley that cuts through the southern end of Clearfield Mountain. I made a mental note of the fact that past the tower the railroad down in the valley went into a tunnel.

By now, there was a very high overcast. Under it, the visibility was a good 10 miles. The Cub had no trouble climbing up to about 2,800feet and flying over Clearfield mountain with its radar domes and radio towers. After having flown the Cub over and through the Rocky Mountains, naming this ridge a mountain seemed to be a misnomer.

At Clearfield, the airport was out of 80-octane, but the FBO indicated he would have some the next day. That did not do us a lot of good, so we used 100LL. The A&P that does my annual inspections is a believer in adding Marvel Mystery Oil to the gasoline each time you top off the tank. This practice is the aviation equivalent of an "Urban Myth." Scientifically, it has not been proven that the oil does anything to protect the engine valves, but none of us is willing to take the risk of not using it. Consequently, each time the Cub's tank is topped off with 100LL, I add a couple of ounces of Marvel Mystery Oil. Every so often, I even add it to the 80-octane.

While we were having lunch at Clearfield, a Cub landed with a fellow in his 40's and his 10-year-old son in it. They were on their way from Kentucky to Lock Haven. The father asked:

"Where are you fellows headed?"

"We are on our way to the Fly-in at Lock Haven."

"Do you mind, if we fly along with you?"

" Not at all, that would be enjoyable."

"My son and I will follow your Cub."

After takeoff, I looked around for him. He was a 1,000 feet above us and about 2 miles south of our course.

By now, I was quite familiar with this last 52 miles. The line of flight was a few miles north of the village of Snowshoe that is alongside Interstate 80. Since Bernie and I were looking up at this village, we concluded that the straight across route is a few hundred feet lower than the Interstate route. Bernie and I were flying at about 400feet, an altitude that would allow us to clear the last ridge before the Lock Haven airport. The pucker factor, although still there, seemed to be a bit less than previous trips. Right on schedule, we flew past Orviston that is down in the bottom of a narrow valley. Two miles further east, in the same valley, is another hamlet that rates a circular symbol on the map but not a name. Bernie and I call these "no-names." To the south, the wide V in the east-west ridge stood out against the horizon.

Letting down into Lock Haven the other Cub rejoined us. The two Cub's entered the right-hand traffic pattern and landed, to the west, one after the other on the grass runway. Touch down was at 2:45 P.M. and my Cub's engine hours meter registered another milestone--2,000 hours. Nobody seemed interested in showing us where to park, so I taxied to the end of one of the first rows and parked. The other Cub parked a few rows further to the west.

Bernie and I went to the registration table and started to fill out the paper work. The lady behind the desk asked us where we were staying and I told her the Fallon Hotel, she said, in no uncertain terms, that we did not want to say there. In earlier years, it had been a prime place to stay and I had always felt the Holiday Inn (by now a Comfort Inn) was too far away. I thought staying at the Fallon was a good idea. The lady said a fellow bought it to make a killing on the flood control levee that would take part of the hotel and all of its motel. He had let the place go to ruin and it had acquired an unsavory reputation as a "no tell motel." She called the guy up, after a few stern words canceled the reservation. Then, she found us a room, for two nights, at the Hoffman Bed and Breakfast that was only a half mile or so from the north side of the airport.

As usual, we were only on the ground a few minutes before encountering Dick and Dan Knutson as well as Don and Steve Johnson. I had talked to the Johnsons at Oshkosh where they said they were not going to Lock Haven this year. But, like myself they could not stay away.

After dinner at the pavilion, a van took us to the B&B where Bernie and I were assigned a small room with a bed and roll-away. The bathroom was across the hall--shades of England. To get to the room, we had to walk through the kitchen. The house had once been a large mansion and was decorated in the usual Victorian manner. At breakfast the next morning, the proprietress waited on tables wearing a well-worn house coat and rollers in her hair. It was obvious that she hadn't quite caught onto the fact that B&B's and gracious living go together. Despite this, the pancakes were great. An exchange student from the Ukraine, who was a computer science student, was working for her this summer. He seemed to be quite up to date on hardware and software, but was only interested in mainframe computers. I concluded that he had little or no access to personal computers.

Since our room was only for Thursday and Friday nights, Bernie and I still had the problem of finding a place to stay for Saturday night. The lady at the B&B called the Comfort Inn and found that they had a number of rooms available. Bernie and I quickly reserved a room for Saturday night. The weather to the west of Clearfield Mountain had gone bad. Quite a few people flying to Lock Haven became grounded and had to cancel their reservations. It was apparent that Bernie and I were almost the last ones to fly into Lock Haven from the west. I called Ann and told her about the Fallon problem. She canceled and also reserved a room at the Comfort Inn.

Friday afternoon, I was sitting by my Cub when Clyde Smith, Sr stopped to check the brace wires on the Cub's tail. He plucked them, listened, and then said "just right." He told me that most of the Cubs here had their tail brace wires either too tight or too loose. I didn't tell him that until now I had no idea what the correct tightness was. Late in the day, I discovered that for the second time at Lock Haven, the Cub's electrical switches had been left on. This was despite the warning placard above the left window. I had to pull out the battery and take it to the maintenance shop on the other side of the airport to get charged. The excitement of making it over the solid woods made me forget to attend to the switches again.

On Saturday, Ann was scheduled to arrive at noon. An hour and a half later, I saw a dark blue Camry come speeding up the lane--Ann had arrived. She and I had lunch at the pavilion and then spent a couple of hours walking down the lines of airplanes. There were a lot of nice Cubs, a Pitcairn Mailwing biplane with an OX-5 engine, a rare Doyle parasol, and a Stinson SR-8. Back at the pavilion, we encountered Bob and Diane Dempster. Much to my surprise, Bob repeated the Montana story to Ann and asked her some questions about the incident.

The Dempsters had a new adventure to relate. From Egypt, they had shipped the Cub to Miami, Florida where they left it with his parents. It had been there for the better part of a year. The Dominican Republic had scheduled a Fly-in to attract American tourists. Since their Cub was in that part of the world, the Dempsters decided to attend the Fly-in. They equipped the J-3 Cub with a Garmin GPS, an artificial horizon, a life raft, Mae West's and all the other required survival gear. Diane said that, when they wore all the gear, it felt like they had been stuffed into the Cub. Neither of them could do more than wiggle. I really don't know how they could get all that equipment in the Cub. With 36 gallons of fuel onboard, they were hundreds of pounds over the Cub's allowable gross weight.

Bob and Diane flew from Miami to the Bahamas and then hopped islands to the south. The last flight segment was 150 miles over open water, but the GPS kept them informed as to where they were and how fast they were going. Upon arriving at the Dominican Republic, they had a bit of a hassle with the military over communications difficulties, but they got it sorted out. Later, they followed the same route back to Miami.

I also talked to Len Buckles, from San Diego, who was having a great time. He is retired from the Army and is a frequent contributor to the Cub Club newsletter. This was the third or fourth time he had flown to Lock Haven. The first time he was accompanied by a 75-year-old friend. This time his copilot was a very attractive young lady who had ambitions of becoming an airline pilot. Needing to build up some flying time, she decided that a slow trip to Lock Haven would be a good way to do it. Len was envied by the other Cub pilots. He spent quite a bit of time explaining that she was like a daughter to him. I asked Len if he found flying the 5,000 mile round trip difficult to do. His reply was that it simply amounted to a lot of 100 mile hops.

There was another young lady from San Diego, Lisa Cook whose husband had been transferred to Washington, D.C.. She and another woman flew the Cook's Cub from San Diego to D.C.. Then, she and her husband flew it to Lock Haven. Such long flights are quite common at Lock Haven as there were a number of pilots from California, Washington, Canada, and other distant parts. One poor PA-12 pilot fought the weather all the way from the State of Washington only to arrive the afternoon of the last day of the Fly-in.

There were two seventeen year olds, Chris Price and Josh Brown, from Sonoma, California (near San Francisco).They had a Cub with McDonald's golden arches logo and "The Spirit of Sonoma" painted on the fuselage. They had flown their Cub from Sonoma to Tehachapie to Phoenix to El Paso with the temperature at 112 degrees on these last two legs. From there they went to St. Louis and then to the EAA convention at Oshkosh. I saw the Cub at Oshkosh, but at the time I did not know who was flying it. After Oshkosh, they flew the Cub to Old Rinebeck, New York to visit Cole Palen's WW I aircraft museum and from there to Lock Haven. I talked with Josh Brown one afternoon. He said that he now had more cross-country time in the Cub than he had total flying time before he left on the trip. Josh had received his Private Pilot license only three days before they left. Chris had his for only three months. Their navigational technique was to follow the Interstate highways that were high-lighted in yellow on their maps. Both of them were very unpretentious and were having the adventure of a lifetime. In the summer of 2000, Chris Price got a job flying BAC 146 commuter airliners for Air Wisconsin and now keeps the Cub in his hangar at Brodhead, Wisconsin. Later, Josh got Chris's old job of flying DC-3's out of Middletown, Ohio.

Saturday night, Bernie, Ann and I attended the banquet. It was held in a very large tent that was a major improvement over the open pavilion of previous years. After the banquet, Ann drove us to the Comfort Inn.

The weather Sunday morning was miserable: cold, windy, and white scud clouds were drifting through the valley. A solid cloud deck hung halfway up the big ridge to the south of the airport and it showed no sign of lifting. Having previously endured rain at Lock Haven, I had insisted that Bernie and I bring rain gear along on this trip. However, I was not prepared for the cold. Ann let me borrow the Piper Museum sweatshirt she had bought. It kept me warm under the rain jacket.

The three of us went to Chillcot's for breakfast, so Ann could see some of the local color and characters. This restaurant is in a small house across the alley from the west end of the old Piper plant. The restaurant has been in operation since the 1930's. In those days, it was popular with the Piper employees. As the small tables were very close together, the inside of the restaurant was very crowded. There also were five seats at the counter and a small kitchen. The wife of the proprietor was more than 80 years old, but had the complexion of a young child. Ann was impressed with this small family-operated restaurant that hadn't changed one iota over the years. When it became obvious that the Cub was not going anywhere today, Ann headed back to Wilmington. Bernie and I reserved a room at the Days Inn, near downtown, and had a place to stay for another night. Then, we spent the rest of the day wandering around the Fly-in grounds while trying to stay out of the light rain and drizzle.

During the day, a steady trickle of planes moved from the Fly-in side of the airport to the hard stand on the north side of the runway. It took me a while to catch on, but these were pilots from the East Coast who had to get to work on Monday. They were leaving their planes at Lock Haven and getting automobile rides back home. By the end of the day, the south side of the airport was almost deserted. Only a scattering of planes were still parked there. By mid-afternoon, Bern and I packed it in and went to the motel.

Monday morning the weather was a bit better, but the solid cloud deck was only a couple of a hundred feet or so above the tops of the ridges. The visibility was only a mile or two with patches of fog in the valley. As we entered the airport, a solitary Cub following the Interstate 80 to the west disappeared into the gloom.

When the gas truck came to the Cub, I had him top off the tank. It took only nine tenths of a gallon. When the driver went to move the truck, it would not start. Then a second truck, with the jumper cables, came and it got stuck in the wet ground. Later, I heard later that the FBO was really angry about all the trouble nine tenths of a gallon caused him. By 11:00 A.M. a couple of planes took off and headed north, but the bulk of the remaining pilots just watched the weather. By this time, I decided that we had better leave before we were stuck here for a long time. While Bernie packed up the Cub, I told Steve and Don Johnson that we were leaving.

The Cub, Bernie, and I took off to the west and slowly climbed up to the height of the first ridge to the west of the airport. We had decided to take the straight across route again. Approaching the ridge, I was pleasantly surprised to find 3-4 mile visibility under a 300 foot ceiling. Once over the first ridge, we were able to drop down another 50-100 feet and had excellent "Piper Cub" flying conditions. One of the important landmarks on the west end of this segment is a blue water tower at an unnamed collection of houses. If you are close to it, you are north of course, but past the completely wooded area. Bernie did a nice job of map reading and the Cub stayed exactly on course to Clearfield. After landing, I taxied the Cub up to the gas pit where there was a Cub, with Civil Air Patrol markings, that belonged to Bill Davis from the Detroit area. It was the Cub we had seen leaving around 8 A.M..

Bernie and I went to the restaurant to have lunch. Just as we sat down at a table, the telephone rang. The cook answered it, then she yelled out into the room "Is Frank here?" When I got over the surprise, I picked up the phone, Steve Johnson was on the other end of the line. He wanted to know about the weather. I told him the visibility was 3-4 miles under a 400foot ceiling with no wind. About the time we finished talking, the two kids from Sonoma landed and came in for lunch.

Clearfield Mountain to the west was obscured by clouds to a point just above the level of the airport. In the valley below the clouds the visibility was at least five miles in haze. Both Davis and the kids wanted to go northwest toward Detroit, but were stymied by the low clouds. I told them that I was planning to go around the south end of the ridge to Punxsutawney. When I showed them the route on the map, I also pointed out the valley with the railroad that I had followed through the ridge the previous year. After some discussion, it was decided that I would lead them to the valley and through the ridge. I told them there was a radio tower near the northwest end of the valley, when they saw it to make a left turn and they would be on the west side of Clearfield Mountain.

The kids took off first, followed by Bernie and me, and Bill Davis was last. The group of three Cubs headed southwest past the city of Clearfield with myself in the lead. After flying nine miles, the mouth of the railroad valley appeared. My map showed a highway just to the west of the valley and I decided to follow it. The kids did not seem to like this. They peeled off and dove down into the valley. It was a beautiful sight as their bright yellow Cub seemed almost spotlighted against the dark green walls of the valley. The thick black and grey clouds, just above the top of the ridges, were spewing scattered columns of rain that swept through the tree tops. Because of the scale of things, the kid's Cub looked just like a yellow Canary against the background. Bill Davis apparently liked the valley route as he had formed up behind the kids.

Bernie and I went west a little further, but I could not find the highway. I turned the Cub toward the valley just in time to see the kids and Bill Davis disappear into the haze ahead of us. I hoped they would not fly too low in the valley. I had forgotten to tell them that, at the northwest end of the valley, the railroad enters a tunnel. Bernie and I flew along the west side of the valley and finally found the missing highway. About the same time that I saw the radio tower, the kid's Cub peeled off from somewhere to our right and passed between us and the tower. Upon reaching the tower, they made a pylon turn around it and flew off to the side. In a minute or so, they pulled up close to us, looked us over, and decided that we were not the Cub going to Detroit. About a mile to our northeast, Bill Davis's Cub could be seen despite the haze. The kids must have seen it too as they cut across in front of us again. In a few minutes, they had positioned themselves well behind Bill. Those two Cubs slowly pulled away from us toward the northwest and finally disappeared into the grey haze.

Some twenty miles west of Clearfield Mountain, the ceiling went up to an estimated 4,000feet. The visibility was a good 10 miles in light haze. Bernie kept us on course as we headed for the Youngstown-Elser airport. As the Cub approached New Castle, I happened to look up and saw a Piper Pacer go over us headed due west. I told Bernie that it looked like Steve Johnson's airplane. By the time we reached the Youngstown-Elser airport, the bend in the wire fuel gauge was an inch or so above the cap. Having plenty of fuel, Bernie and I decided to fly on to Salem only eleven miles further. In addition, I did not really want to takeoff to the west again at the Youngstown-Elser airport.

As the Cub entered the Salem traffic pattern, I could see three Pipers (two Pacers and a Tri-Pacer) parked at the gas pit. They were Steve Johnson's group. I landed, taxied the Cub up to the pit, shut down, and Bernie and I climbed out. The first thing I asked Steve was:

" How did you like the weather?"

He replied:

"It was just as you said."

Mike, in the second Pacer said:

"It was a bit dicey, but flyable."

Tim, the Tri-Pacer pilot said:

"It was the worst I have ever seen and I don't want to do that again, ever!"

Some eight years later, I talked to Tim at the Grassroots Fly-in. He remembered the flight vividly and recalled talking with me at Salem.

Bernie and I went to start the Cub. Bernie climbed in the front seat, I pulled the prop through and nothing happened. After about 10 minutes of going through the usual hard starting regimes we gave up. Now, it was a question of dirty fuel (as per the last trip) or overheated magnetos. After some discussion, we decided it was the latter. Bernie went into the airport office to get some ice and water, I took off the upper cowl and positioned the Cub so that the breeze could blow across the engine. Bernie kept pouring cold water on the mags until they were only warm to the touch. Unfortunately, the engine still would not even pop. Finally, a Cessna 182 taxied out, did his run up about 50 yards away with the prop blast aimed right at the Cub's engine. Ordinarily, I would not have appreciated this, but today it was just what we needed. After this, the mags were cold to the touch. Bernie and I buttoned up the Cub's cowl and the engine started at the first pull on the prop. Without wasting a second, we took off for Ashland. We landed there around 5:30 P.M. with enough daylight left to make it to Findley or even a bit further. It was not to be as the Cub's engine would not start. By the time it cooled down we had run out of daylight and had to RON at a motel.

Eight o'clock the next morning(Tuesday), the weather was perfect. Bernie wiped the dew off the windshield while I loaded our gear into the Cub. The engine started with the first flip and we took off for Findley instead of Priebe. After refueling the Cub at Findley, we had to go through the same drill to cool off the magnetos-- another hour wasted. Bernie was the pilot from Findlay to Warsaw. By now, he was really good at holding a heading, keeping on course, and maintaining altitude. Since we were flying over familiar country, I had him fly lower than we had on the trip out, but it did not make any difference to him. Reaching Warsaw, I gave a call on the radio, entered the traffic pattern and landed to the west.

The fellow who managed the airport office at Warsaw was bit of a prophet of doom, who must only fly in nice weather. He said there was a line of thunderstorms to the west that we could not get through. I thanked him for the advice. After lunch at Grandpa's restaurant again, we poured more cold water on the engine. Bernie and I were getting quite good at knowing at what temperature we should try to start the engine. Then, with myself as pilot, we took off took off and headed for Joliet, Illinois. As we flew past Valparaiso, under a medium height grey ceiling, the Cub encountered a strong head wind. Bernie calculated a 45mph ground speed. From the position of the gas gauge wire, it was clear that we did not have enough fuel to reach Joliet. I knew there were several airports on the course line. We could land at one of them when we needed to do so.

After two hours in the air, the Howell-New Lenox, Illinois airport appeared off the nose. Since we were now flying on the reserve fuel, we landed there for gas. On the ground, there was a 20 mph wind and it was quite cold for August. We refueled the Cub, took off the top engine cowl, and positioned the Cub so the wind could cool the engine. Every 10 minutes, the Cub was rotated about 45 degrees, so the other side of the engine would feel the impact of the wind. In about 30 minutes, the engine felt cold enough, Bernie and I put the cowl on, and the engine started on the first flip.

From here to Joliet, the line of flight is over the Chicago suburbs. There are many large office complexes, shopping malls, miles of housing, four lane highways, railroad tracks, and numerous power lines. I do not like flying over this area as there are very few places to land if the Cub has problems. As we passed over the Joliet airport, the Cub broke out into clear skies, a warm sun, and no wind, so much for the line of thunderstorms. West of Joliet the countryside is mostly farmland and it is much more comfortable to fly over than the suburbs. Since we had plenty of fuel, Bernie and I flew to Plano and then turned north to Harvard. In the 1960's this airport was one of the most active airports in the Midwest. Many airline pilots kept their Stearmans, Cubs, and other airplanes here. Now it is practically deserted and all the facilities are badly run down.

Bernie and I landed there about 5:30 P.M.. I was concerned that they might have closed for the day. As we taxied the Cub up to the office building, I saw a woman. I shut down the engine, climbed out, and told her that we needed fuel. She said that she would tell her father and she drove off. About 10 minutes later her father's truck went along the row of hangars and out the driveway. Bernie and I thought he was leaving, but he came back into the airport through a second entrance. Due to EPA regulations, their underground gasoline tank had been removed. An old tanker truck was being used for fuel storage. These regulations were disastrous for the small owner operated airports. At Waunakee, Jerome Ripp spent $27,000 removing the underground tanks and replacing them with above ground tanks. The State of Wisconsin had a program that was supposed to reimburse him, but it ran out of funds. I never did find out if he ever received any reimbursement.

From Harvard to Madison the weather was perfect. There was no wind and I did not even need a map, but I let Bern keep on navigating just to keep us honest. Finally, the state capitol came into view. After flying through Madison's nest of tall television towers, it was 7:14 P.M. when the Cub landed at Waunakee. The total round trip flying time was 24 hours 15 minutes yielding an average ground speed of 57mph. Another Sentimental Journey was history.

Once home, I also discovered that the rotors inside the magneto's were wobbling. I installed a set of used magneto coils, new shafts and ball bearings. After this, the engine started easily and ran very smoothly. As I would eventually find out, these repairs were not to be the long term solution to the hot mags problem.

Postscript

Between 1989 and 2006, the Cub and I have been to Lock Haven ten times over basically the same "80-octane" route. The weather has varied from beautiful to a grey cocoon with rain and limited visibility. The average ground speed has ranged from 57 to 66mph.On two occasions, the Cub made it from Lock Haven to Waunakee in one day. Once east bound, I had to turn back at Fort Wayne, Indiana when I ran out of ceiling, visibility, and nerve all at the same time. I never made it to Lock Haven(the trip did not count).

The Sentimental Journey Fly-in's at Lock Haven always are enjoyable. At the early ones it was very interesting to talk to the old men who had built the Cubs. At all of these Fly-ins, the best part is talking to Cub pilots from all over the United States. Every one of them thought nothing of strapping on a Cub, flying for hundreds or even thousands of miles across many different kinds of terrain, and through all kinds of weather to reach Lock Haven. Although the Cub is not a very comfortable plane to sit in for days-on-end, they enjoyed every minute of their flights.

For a number of years, the Fly-in had suffered from poor weather in August that held down attendance. It also generated considerable comments from people who had been weathered-in while en route. From 1997 on, to try to get better weather, the Fly-in was scheduled for the third week in June. Unfortunately, this new schedule did little to improve the weather situation. In both 2001 and 2003 the Fly-in was plagued by low clouds, rain, and some fog for all three days. In contrast, 2002 was blessed by beautiful weather--a year I did not go.

This schizophrenic weather is a function of Lock Haven's location. The Susquehanna River comes out of the mountains at the northwest edge of Lock Haven and enters a valley that becomes wider as it goes east. The small end of this valley ends at the ridge just west of Lock Haven. When a strong low pressure system positions itself on the Atlantic Ocean coastline, the counterclockwise east wind pushes cold, moist air from the ocean into the valley. The result is the low clouds, rain, and fog that is common at Lock Haven. In addition to the low cloud deck there are wisps of white clouds that reach to the ground. They seem to attach themselves to the tree tops and not move. Because the air is quite cold, it takes a considerable amount of heat from the sun to cause the clouds to rise. As a result, once these conditions establish themselves, it can be several days before flyable weather returns. None-the-less the faithful return to Lock Haven each year. Being one of them, there are Sentimental Journeys yet to be flown.

Postscript

Chapter 16
To Denver-Twice

The First Trip–1994

In the spring of 1994, my brother John suggested I fly the Cub to Denver to go with him on a white water rafting trip on the Arkansas River. Since Bernie had enjoyed the 1992 trip to Lock Haven, I asked him if he would like to go on this trip. He agreed to do so as he had never been white water rafting. In that I had been to Denver on the "southern route" a number of times, I laid out a new route to North Platte that went across northern Iowa and then to Norfolk, Nebraska. From there the route went on a southwest diagonal across the prairie to North Platte.

The Cub, Bernie and I headed west from Waunakee early on Friday, the first of June 1994. The weather was pleasant, but there was a slight head wind. Just past Decorah, Iowa the Cub was flying at the same speed as a beat-up farm truck on the road below. The driver had his arm resting on the window sill and was looking up at us. Bernie noticed him and started watching the truck. This joint admiration society went on for a few miles until the truck finally turned off the highway. We both wondered what the truck driver thought of the low flying yellow airplane that could go no faster than his rather tired looking truck.

Bernie and I made fuel stops at Clarion and Storm Lake, Iowa as well as Norfolk, Nebraska. From Norfolk to Broken Bow the normally brown hills were green with the spring's new grass. There were occasional trees with freshly minted leaves. Amidst all this were widely dispersed cattle. There were many calves staying close to their mothers. Other than the ubiquitous windmills there were very few signs of civilization.

When Bill Vogelsang and I had approached Broken Bow, a few years earlier, it was from the northwest. Now, Bernie and I were coming in from the northeast and there were a totally new set of checkpoints. Eighty miles southwest of Norfolk we crossed the North Loup River at the town of Ord. It is on the west side of the river tucked between the water and a bluff. Twenty miles further, we flew over the village of Comstock that was on the east side of the Middle Loup River.

In that it now was late in the day, I was becoming worried about the availability of fuel at Broken Bow. As the Cub entered the traffic pattern, Bernie reported seeing people at the airport. The Cub touched down at 4:54 P.M. and was taxied up to the fuel pumps. The regulation bound ex-sergeant had been replaced by a young fellow associated with a crop spraying outfit. He did not know the undertaker, who had been a P-38 pilot.

The leg from Broken Bow to North Platte was over desolate prairie. Only a few roads were depicted on the map and there were no symbols for ranches. A few miles west of Broken Bow there is a very wide, but deep valley having a north-south orientation. The western side of the valley consists of a series of grass covered rounded swells. A few isolated trees dotted the swells, but there were no signs of civilization in the valley.

A tall power line, three miles to the south of our course line, went directly from Broken Bow to North Platte. In the receding light, its pylons stood out against the horizon. They took the stress out of navigating across this sparsely populated country. I had flown this leg, in the opposite direction, on a previous Denver trip, but had not seen this power line before. I concluded that it was new. As is usually the case on these desolate flight segments, we never saw any people or even a vehicle on a road.

After landing at North Platte at 6:10 P.M., the weather to the west of there looked as if it might rain. To be safe, the Cub was put in a hangar for the night. While conversing with the FBO, I told him about the Cub spending three weeks here in the late 1960's. He said that he was a teenager working at the airport at the time and remembered the Cub.

Bernie and I stayed overnight at the Stockman's Inn as it provides transportation to and from the airport. Unfortunately, their kitchen was closed due to a fire in it the previous week, so we had to find another place to have dinner. North Platte is a major stopping point for travelers on Interstate 80. As a result, it has a large number of motels and restaurants lining the broad street leading from the downtown area to the Interstate. We just picked the closest restaurant.

The next morning we awoke to find a low ceiling hanging over the city. Once at the airport, Bernie and I watched the clouds for a half hour. The weather seemed marginal, but stable. Since there no longer was a fixed base operation at the Julesburg airport, a different fuel stop was needed. We considered going to Sterling, Colorado, but the 136 mile leg was too long given the present weather conditions. Ogallala, Nebraska, only 54 miles to the west, would be our next fuel stop.

As the Cub broke ground, I made a tight left turn to the west and at an altitude of 200feet started following the railroad tracks to Ogallala. On the west side of North Platte, there is a large railroad switching complex with many parallel tracks. Long lines of box cars were being shunted to the various tracks in the yards by men in a control tower. Just past the yards, I drifted the Cub over to the south side of the railroad tracks. We were flying just under the clouds with visibility of a mile. Half way to Ogallala, Bernie and I saw a white and blue Piper Super Cub that suddenly appeared on the north side of the tracks. The plane was headed east about 100 feet lower than we were--shades of Dodge Center. In a few seconds, it disappeared into the haze.

Ogallala has a long east-west runway and a shorter cross runway that goes up a hill to the northwest. The hangars, office building and the fuel pumps were at the east end of the airport. In order to minimize the distance I had to taxi, I landed the Cub to the east. As we refueled the Cub, the line boy was asked about the Super Cub that Bernie and I had seen. He said that it was engaged in pipeline patrol and stopped here every couple of weeks. The line boy also mentioned that he was restoring a Cub and needed some drawings for it. I told him that the Cub Club had many of the engineering drawings. After returning home I sent him the club's mailing address.

As the Cub flew toward Sterling, the weather improved. By the time we refueled there, the clouds had dissipated. From Sterling to Denver the ground gains another 1,000feet in altitude, so the Cub was climbing along the whole leg. Two miles northeast of Denver's Tri-County airport there are two very tall television towers that form a line pointing directly to the airport. Despite this, we had some trouble finding the airport amidst all the ground clutter of buildings and roads. John was waiting for us and after we landed at Tri-County, the Cub was taxied over to John's hangar. After much pushing and shoving of airplanes, the three of us finally wedged the Cub inside the hangar.

After dinner, John, Joye, Bernie, and I loaded all of the rafting equipment into the trailer. The next morning, John's son-in-law, Ed Lafferty, joined us. Then, the five of us drove to the Arkansas River in the mountains west of Canon City, Colorado. As it was still snow melt time, the river was running at 3,200 cubic feet per second that is much higher than normal. While the four of us men ran the river, Joye drove the van and stopped at each pull-off to watch our progress.

On this trip, John had configured his two tube cataraft as a paddle boat. Ed and I were each riding near the front of a tube as if we were on horseback. John and Bernie did the same near the rear of the tubes. Since Bernie had never been on a white water rafting expedition, John wanted him nearby so he could instruct him in handling the raft. We practiced controlling the raft with our paddles before entering the narrow Three Points rapids that is rated as class 3-4. Making it through in good shape, we then went down long stretches of river at a good clip. The last big class 3-4 rapids, called Five Points, was negotiated with no trouble.

About a mile below Five Points there was a relatively straight section of river with a line of what are called haystacks in the center of the river. These result from hidden rocks causing the water to form a 3-4 foot mound of water in front of the rock, i.e a haystack. Behind the rock there is a deep depression in the water. For bit more excitement, John yelled at us to hit the haystack's head on. The first two were ridden over in good shape. When we went into the third one, there was a huge deep hole behind the rock. The raft went down into it at a 45 degree angle. When the raft hit bottom, I suddenly found myself flying through the air and into the raging water. I quickly put my feet out in front of me to fend off any rocks hidden in the water. My arms were put out to the side to stabilize myself.

In a minute or so Ed floated past me. I was so surprised to see him that all I could say was "Ed, you are in the water too." Looking back at the raft, the only person still on board was Bernie who had no idea as to what to do next. Fortunately, when John went overboard, he grabbed a rope tied to a pontoon and was able to get back onto the raft. Then, he and Bernie maneuvered the raft and picked up Ed.

I tried to grab onto some brush on the north side of the river, but was moving too fast to hold on. I went back toward the edge of the haystacks and floated down the river with only my Tilley hat showing above the water. Under the hat, my face was out of the water, I could see where I was going, and I was riding nicely downstream. In a minute or so, Ed reached down, grabbed my life vest, and pulled me back onto the raft. Much to my surprise, the raft caught up with the paddles that had flown out of our hands. We snatched them out of the water and started paddling again. Pulling up to shore, the raft was tied to a rock in order to recoup after our misadventure. John said that he thought I had drowned as only my hat could be seen floating on the water.

While we were resting, several commercial rafts hit the same haystack and flipped over. There were people and paddles floating everywhere. Anticipating this event, the companies had two men in kayaks downstream catching paddles and throwing them up on the south river bank for later retrieval. Joye took several photographs of our adventure in the haystacks. At the take-out beach, John declared that Bernie now was a genuine river rat.

Bernie and I had enough excitement for a while, so the next day (Wednesday the 6'th of June) we headed home. The route of flight was to Cheyenne and then to Scottsbluff. Instead of going to Rapid City, South Dakota, I planned to go to Gordon, Nebraska and then fly east parallel to Nebraska's northern border to the Missouri River. I had used this route on a previous Denver trip.

There is a hill at the south end of the runway at the Tri-County airport. After breaking ground, the Cub cleared the hilltop by only a few feet. From where John was standing, it looked as if the Cub was not going to get over the hill. I made a right hand turn to a northerly heading and then bid John adieu over the radio.

The route of flight took us first to Cheyenne, Wyoming where, despite the 6,156 foot altitude and a control tower, we had no trouble landing or taking off. From there to Scottsbluff, Nebraska, we would fly what was a familiar route for me, but it was new to Bernie. As we left Cheyenne, on a compass heading of 40 degrees, he got time hack at a good check point and a few miles further another one at a power line. It was the last checkpoint he was to see for quite a while. Finally, the lone microwave tower, with its adjacent crossing power lines, appeared right at Bernie's ETA. Five miles past the tower the ground takes a definite downhill slant to the northeast for the next 30 miles. On an earlier Denver trip, I had flown this same leg in the opposite direction and was climbing up this slope the whole way. Now, Bernie and I were going downhill over this same track. The ground speed on this leg, aided by a good tailwind, reached an astounding 95 mph. A speed rarely attained in the Cub.

The few signs of civilization were a couple of roads, a ranch that was depicted on the map, and a power line perpendicular to our line of flight. Just before Scottsbluff there was a high ridge directly in front of us. At the east end of this ridge were a half-dozen sharply pointed hills. These gave this part of the ridge the appearance of a distant mountain range. Once over the top of this ridge, Bernie and I could see the famous bluffs. The Cub touched down at Scottsbluff at 10:22 A.M. The FSS at Scottsbluff had been closed. No great loss, as I never had been able to get them to answer my radio calls.

After taking off from Scottsbluff, Bernie and I had to thread our way between a pair of wildlife refuges where the minimum over-flight altitude was 2,000feet AGL.. An altitude that the Cub rarely reaches. Strangely, I had not noticed them on my previous flights along this leg. Perhaps they were new. From there to Chadron, it was an easy 70 mile flight over the almost uninhabited prairie. There were only a few roads and ranches for check points. A few miles southwest of the Chadron airport, there is broad valley with a creek in the middle. The sides of the valley slope upwards towards the gently rolling prairie. There were a number of Cottonwood trees scattered around the slopes. Near the creek the grass had its spring time color of lush green. The scenery here deserved a second look, but we continued on course to a fuel stop at Chadron.

From Chadron to Gordon it was only a forty-five mile leg on a heading of 82 degrees. This is a much shorter leg than normal, but the next airport beyond it (Valentine) was too far away. As the Cub approached Gordon, the afternoon sun disappeared behind a dark grey cloud shield. Bernie and I could see thunderstorms to our south and to our north that were not there earlier in the day. After my experience at LaCrosse with Ann, I doubted that we could outrun them.

When lines of thunderstorms appear in the Midwest, they usually develop in conjunction with fronts that take a day or two to march off to the east. Typically, there is enough advance warning of the storms to plan accordingly. In the Great Plains, it can be a perfectly clear day and in a few hours gigantic towering thunderstorms can develop seemingly out of nowhere. The Cub and I have had the displeasure of encountering a number of them over the years. I suspect that the sudden development of such storms results from moist air coming up from the Gulf of Mexico and colliding with cooler air spilling down over the Rocky Mountains. Today, Bernie and I might have to deal with yet another collision.

We made it to Gordon with the line of thunderstorms nibbling at our backs. While refueling the Cub, the Gordon airport manager told us that one of the local planes was on an extended trip, so its hangar was vacant. He indicated that we were welcome to put the Cub inside the hangar for the night. The three of us tucked the Cub away in the hangar. Then, the manger gave us the keys to the courtesy car. Bernie and I drove to a downtown motel where we checked in. The line of thunderstorms hit Gordon an hour and a half later. The considerable lightning and thunder were accompanied by heavy rain and strong winds. We were grateful that the Cub was in a hangar.

The next morning, it was rather warm with a strong wind blowing from the south. In order to get to Valentine, Nebraska we had to get the Cub over a low northwest-southeast ridge just to the east of Gordon. I started climbing on course and after a few miles, it became clear that the Cub was not going to clear the ridge. Bernie, the navigator, said that the ridge was lower on the northeast side, so I deviated from the course in that direction. As the Cub went past the north end of the ridge, the down spilling air held us in its grip. The Cub barely held its altitude. As Bernie predicted, the level of the northeast rim was slightly lower than the main top of the ridge. The Cub skimmed over the ground with its wheels only a couple of feet above the rocks sticking out of the tall grass. After a few hundred yards of this we finally flew out over lower ground and were on our way to Valentine.

The line of flight from Gordon to Valentine basically follows the Niobrara River. A power line connecting the two towns was at a slight angle to our course line and crossed it near the midpoint. I did not recall seeing this power line on a previous trip. Either it was new or the visibility had been limited in the past. The countryside here is prairie with very few signs of civilization. Twenty eight miles east of Gordon, the Cub crossed the Niobrara River where it flowed through a cliff lined gorge. The south facing cliffs were bare dirt while small trees clung to the north facing cliffs. This area is one of the prettiest in all of Nebraska. It is basically rolling prairie with an occasional dirt road, a few cottonwood trees here and there, with the Niobrara River slicing through the countryside. At one point a pickup truck, going north and leaving a plume of dust, went over a bridge crossing a small dry creek bed. The truck appeared to have come from one of two ranches depicted on the map that were two miles south of the river.

The Niobrara River gorge was crossed again where the power line went across the river. There was a ranch house perched on the top of the south side of the gorge. At this point, the gorge ended and the river entered a broad valley that went to the southeast. Our line of flight now took us away from the river. However, Bernie was able to keep the power line in sight the rest of the way to Valentine.

At Valentine, I asked the FBO about the young fellow who ran the airport the last time I had landed here. He said that a local banker had insisted upon a charter flight in very bad weather. As the young pilot took off in a Piper Cherokee Six, it veered off the runway into a parked truck, killing all on board.

From Valentine we flew across more desolate prairie to O'Neill, Nebraska for a fuel stop. Then, to Vermilion, South Dakota that is just east of the Missouri River. The airport is on the river flats and the town is on higher ground. The State University of South Dakota is located here. As the city was rather small and in a far corner of the state, It seemed to be an unusual location for the school

At the airport, there were a pair of four engine De Havilland Herons parked on the ramp. They were slowly deteriorating in the harsh climate. The FBO's wife said that he had bought them on speculation. Once, I had flown in a commuter airline's Heron from Phoenix, Arizona to Prescott to visit Ann at her college. A few weeks later, one of the airline's Herons shed a wing in flight.

I had stayed over night in Vermilion once before, but the motels were several miles from the airport and there was no taxi service. Consequently, Bernie and I chose to fly to Storm Lake, Iowa to RON there. A few miles after takeoff from Vermilion, the Cub crossed a small river. In 1993, a stationary front had parked itself horizontally across Iowa and it rained for weeks. Shortly afterwards, I had flown across this same river. Only then it was more than a mile across and had flooded a wide expanse of the valley

After landing at Storm Lake, the Cub was put in a hangar and Bernie called the closest motel. A fellow from the motel, run by a couple from Bombay, India, came out and took us to the motel.

The next morning, the takeoff from Storm Lake was interesting as a major highway is right at the end of runway 30. I made an intersection takeoff from about the midpoint of the long runway. Just about time the Cub broke ground, a semi-tractor trailer came roaring down the highway. Whether or not the Cub would clear the truck appeared to be problematic. None-the-less, the Cub cleared the semi with room to spare.

The rest of the trip home was via Clarion, Charles City, and Decorah, Iowa. At each airport, Bernie and I also alternated duties to offer some variety to the long day of flying. Common wisdom is that flying to the west is into the wind and flying east is with the wind. Again, the Cub defied this logic as our average ground speed to Denver was 69 mph and on the return trip it was only 51 mph, despite one leg being flown at 95 mph.

Second Trip--1996

Two years later John invited Bernie and me to go on yet another white water rafting trip on the Arkansas river. The Cub, Bernie and I departed Waunakee on Sunday, August 11, 1996 via a modified route across Iowa. This time we went through Monticello, Marshalltown and then to Audubon where the town sits on a hillside overlooking the airport. The airport owner lived in a house not far from the gas pit. He was getting on in years. Each time I stopped here, the facilities looked a bit more run down.

From Audubon the route of flight was to Columbus, Nebraska where we joined the original "southern" route to Denver. North Platte is 650 miles from Madison, but for some reason the Cub almost always gets there in one day. The trip often takes twelve to thirteen hours on the clock. Bernie and I stayed over night at the Stockman's Inn again, but this time the dining room was open.

Monday morning, dawned as a nice day for flying with clear skies and little wind. After takeoff, Bernie and I followed the railroad to Ogallala again. There was yet a different fellow running the airport. Even though there was a credit card gas pump, he came out and filled the Cub's tank for us. From there to Sterling it is a routine flight as there are numerous check points along the South Platte River and Interstate 76. Flying several miles to the west of the river, I had the impression of looking down at it even though we were only at a few hundred feet AGL.

As the Cub approached Sterling, Bernie and I could sense that the density altitude was rising. The airport is at 4,038feet and the main runway goes downhill to the southeast. As a result, of the high density altitude, our landing speed was higher than normal and the landing rollout on the downhill runway was very long. I used the last turn off and had to taxi the Cub a long way uphill to the ramp.

By the time we took off, the temperature was in the mid-nineties, and the Cub rolled a long way before getting airborne. At about half the distance to the Tri-County airport, the Cub was flying at 4,700feet, but not gaining altitude. In that this next airport was at 5,130feet, we had a problem. I told Bernie "we need to find a hawk." Approaching the South Platte river, I spotted a hawk circling inside a thermal. Pushing the throttle to full power, I entered a climbing left turn and positioned the Cub just behind the hawk, who did not even bother to look back at us. The Cub was so close to the hawk that I could see the white feathers on the bottom side of his tail. In the strong thermal, the Cub was gaining altitude rapidly. After following the hawk around three or four circuits, the Cub was at 6,000feet. At this altitude the thermal seemed to weaken. As the Cub flew out of the thermal, I lost sight of the hawk.

Since some altitude was lost after exiting the thermal, I flew back to where I thought it might be. As we entered the thermal again, the Cub's left wingtip almost hit the hawk. That got his attention and he quickly flapped away from us. Regaining the lost altitude, I put the Cub back on course.

Coming in to land at the Tri-County airport, we seemed to be sinking much faster than normal. The Cub just plunked down on the runway. Taxiing up to the gas pit, Bernie told me that the thermometer on the office building read 102 degrees. I was afraid to ask about the density altitude. At the gas pit, I checked the oil and again found that the level was down a quart. I was using 80 weight oil on this trip and due to the high temperatures the engine was burning a quart every six hours. The engine's normal consumption is about a quart every 13 hours.

The Cub was tied down on the ramp to a ground cable. To do so, always makes me nervous, as this time of year, sudden severe thunderstorms are quite common in the Denver area. Years ago, the Cub had ridden out 80 mph winds at JEFFCO while tied to a similar cable. The tie-down ring had slid down the left front wing strut and kinked it. I did not want a repeat of such an event.

The Arkansas River was reported as running at only 600 cubic feet per second, which is very slow. John decided that using his cataraft would be too tame. Instead, we would run the river riding on air mattresses. To ride an air mattress, you lay on your back on it and wrap your legs around the front part of the mattress. This limits your forward visibility, but your arms can be used to steer a bit. It seems risky, but in the early 1960's, our mother had gone with John and me down the Arkansas River floating on air mattresses.

The next morning, John, Bernie and I rode in John's van from Denver to the put-in point on the Arkansas River some 15 miles west of Canyon City. Since Joye was not with us to drive the van down stream along the highway, the three of us would have to take turns at driving the van.

As Bernie and I went down the river, we would be passed by commercial rafts with 5-6 tourists and a professional boatman onboard. As they went by, I would tell them in a Scandinavian accent:

"We are on Ole's budget tour. It is only $9.95 per day, but he doesn't serve lunch."

The tourists, who were paying more than $50 apiece, were not amused, and gave us dirty looks. The boatmen thought it was hilarious. They would hover their rafts downstream of each small rapids to watch us float through.

As I went through one such rapids, I was tossed off my air mattress, but was able to grab it and swim to some rocks near the shore. As Bernie approached the rapids, a female boatman yelled at him:

"Don't go in there!"

He yelled back:

" How in the Hell can I not go in there?"

Then, the rapids promptly flipped him off his air mattress. As Bernie went floating by, I grabbed a hold of him and his air mattress. Then, we struggled to shore.

Bernie and I walked around the next big rapids, called Five Points. John joined us and decided that he would ride through this level 3-4 rapids on his air mattress. I took several very nice photos of the event. In one of them, all that was showing, out of the rapids, was John's hat and the upturned end of his air mattress. A woman who worked for one of the commercial rafting companies watched the whole event. She said:

" I have been on the river for 15 years and that is the wildest thing I have ever seen."

Then, Bernie and I floated down the last few miles to the takeout spot. Getting out of the water, we finally realized how cold we really were. After peeling off our wet T shirts, it took more than 30 minutes before our bodies were at some semblance of normal temperature. When everybody was warm and dry, we changed clothes and then went back to Denver in the van.

The next day, Friday, the 16th of August, Bernie and I headed the Cub back home via the original "northern" route pioneered by John and me. Since it still was very hot, Bernie and I decided to go back to Sterling and then go to Scottsbluff. This route would avoid Cheyenne, its 6,156foot altitude, and a severe density altitude problem.

Since the segment from Tri-County to Sterling was one I had flown many times, I did not take a map with me in the front seat. As we neared the end of the flight, I could see a tall building off to the right. To me, it did not seem to be in the proper place. I thought it was the town of Brush. I told Bernie:

"I think that we are off course, that tall building at 2 o'clock should not be there."

Bernie replied: "We are following a highway and a S curve should appear momentarily."

Despite Bernie's assurances, I decided we were not on course. In addition, we were getting low on gas.

When I saw a crop-duster airstrip, I made a quick approach and landed the Cub on a runway that turned out to be composed of fist sized rocks. The Cub's landing gear got a real workout. I taxied up to the gas pit and climbed out. A fellow came out of the hangar and I said:

"My navigator got me lost, can we get some fuel?"

He replied: "Sure, where are you going?"

"Sterling, how far away is it?"

"See that tall building to the northeast?"

"Yes."

"That's Sterling, it is only three miles away."

"Thanks, we do not need fuel after all."

With a sheepish look on my face, I climbed back into the Cub. I had wrongly accused my magnificent navigator of being off course, when he was exactly on course. Bernie showed me where we were on the map and pointed out that after the S curve the highway went straight to Sterling. After this experience, I made sure I followed our normal practice of both the pilot and navigator having maps. Bernie was very nice about my screw-up and did not harass me about it. Although he should have.

It only took a few minutes to fly to Sterling, where 6.8 gallons of fuel were put in the Cub's tank. We were not as low on fuel as I thought. After refueling, we took off and headed the Cub towards Scottsbluff, Nebraska. The compass heading, across the 88 miles of unfamiliar terrain, was 337 degrees (347 degrees minus 10 degrees of east variation). Ten miles out of Sterling, the Cub flew over the North Sterling Reservoir. It was our only checkpoint until we flew over Interstate 80 east of Kimball, Nebraska. From there to Scottsbluff, we flew over prairie where there were only a few roads and power lines to serve as checkpoints. Bernie had become expert at using minor features of the countryside to keep us on course. The map depicted a solitary ranch a few miles off our course line, but at our low altitude we never saw it.

As we were flying over a route that I had not flown before, whether we had enough fuel for the leg was a concern. Thus, our ground speed was of interest as it would determine how much fuel would be used. I watched the Cub's shadow as it went across the prairie and estimated how fast we were going. When Bernie obtained the time at two checkpoints, he would compute the actual ground speed. Usually, my estimate differed from Bernie's number by only a few miles-per-hour.

At Scottsbluff, the Cub's tank was filled. Then, we took off and headed northeast towards Chadron. Although I had flown this leg a number of times, each trip seemed to be a little different. This time, I did not see the pretty valley just southwest of the Chadron airport. In addition, I observed rocks, clumps of trees, and colors of grass that were not seen previously. This seemed to be a function of the direction of takeoff at Scottsbluff and how wide a turn was made before getting established on course. Flying low severely limits how far you can observe things in the rolling prairie, so it is not surprising that different things are seen on each trip.

While flying from Scottsbluff to Chadron, the Cub went over what had been the village of Marsland. The outlines of four barren city blocks could be seen. There were five surviving houses that seemed to be occupied. A few trees were located near these houses. At the northwest corner of Marsland, there is a large two story, white painted, rectangular, wooden building on the top of a low knoll. It looked as if it might have been a church that had lost its steeple. There was a dirt road that led to the front steps of the building. My older sectional maps did not depict Marsland, but the recent maps now show its location.

Landing at Chadron turned out to be a challenge. The wind was at about 20-25 mph and nearly perpendicular to the runway. To compound matters further, the temperature was in the low 90's. A high speed, wing down, one wheel landing was made. It was a bit of a battle to keep everything in concert until the Cub was firmly on the runway. After refueling the Cub, it was tied down on the ramp. While waiting for the wind to quiet down, Bernie and I spent the better part of an hour chatting with the airport locals. They told us that Marsland once was a thriving community. Cowboys from miles around would come to the Saturday night dances. In those days, the town had a reputation for being a very wild place. I neglected to ask if the white building was the dance hall. Finally, Bernie and I concluded that the wind was not going to slow down, so we took off anyway and headed north.

As we approached the Black Hills, Bernie requested that we fly over Mount Rushmore, so he could take pictures of it. I deviated off course and was surprised that it was located so far back in the hills. Since there are commercial helicopter flights along the front of Mount Rushmore, the Park Service has become quite sensitive about low flying airplanes in the area. Because of this, I maintained quite a distance while Bernie took photos. When they were developed, the faces were very small and indistinct.

We made a quick fuel stop at Rapid City, which has a control tower. Then we went on towards Pierre with the usual fuel stop at Philip. Since my last stop in Philip there had been some major improvements. The east-west runway that had been nothing but sagebrush now was a nice grass runway. The north-south runway had been resurfaced with asphalt and was very smooth. In addition, there was a new, younger, well-dressed FBO whose appearance was a major improvement over those of the past. Bernie and I looked over a Cub that he was rebuilding. After refueling our Cub, we took off and headed for Pierre–69 miles away on a compass heading of 60 degrees. Pierre is one of the better places to RON. The Best Western Motel is new and located right on the edge of the Missouri river. They also provide transportation to and from the airport. In addition, there is a good sports bar only a block away that serves excellent sandwiches.

Sunday morning, it was a bright sunny day with little wind. The first flight segment was from Pierre to Huron. Thirty-five miles before Huron, Bernie and I encountered dense ground fog that was a few hundred feet thick. I did not expect to meet ground fog here as the countryside is semiarid. Besides, rain is rare here this time of year. I flew the Cub "on top" hoping that the fog would soon end. Not being able to see the ground, Bernie navigated by radio towers whose tops poked up through the fog. Off to the east, I could not see an end to the fog. Bernie searched the map for an alternative airport. He found one, only ten miles to the north, at Miller, South Dakota. The Cub was deviated from course and we landed there ten minutes later.

The airport was the base for an outfit using Turbo Ag Cat biplanes to spray sunflower fields. The fellow who owned it had started the business flying Piper J-3's as spray-planes. He offered to refuel our Cub if his son and the hired man could get Cub rides. I agreed and took each of them for a big loop around town. Miller was isolated and as far as I could see in any direction there was nothing but brown prairie punctuated by a few scattered fields of sunflowers. In the middle of town was a small pond that the son referred to as the lake. The hired hand seemed more interested in looking for horses than in the airplane ride.

In an hour, we could see that the fog to the south of the airport had dissipated. We thanked the FBO for the fuel, and took off. Huron was bypassed and we landed for gas at Brookings, South Dakota. Bernie and I walked a few city blocks into town and ate lunch at a small ice cream stand across the street from the city park.

Marshall, Minnesota was our last fuel stop before Mankato. This was the airport where Chris and I had landed on the taxiway when the runway was being rebuilt. The airport office was still the same, but now, thanks to the EPA, they had a set of above ground fuel tanks. From Marshall, it was 95 miles to Mankato where we would RON. This latter leg took only an hour and 20 minutes as the Cub had a unusual ground speed of 79 mph.

The next day on the way to La Crosse, we encountered strong headwinds and had to land at Dodge Center, Minnesota for fuel. The radio tower that had given me such a scare on my first solo Cub cross-country was no longer there. The growth of nearby Rochester had made this airport the center of general aviation and homebuilt airplanes for the region. I asked one fellow:

"Do you know Ed Schemerhorn?"
He replied: " You must be that professor from Madison that flies his Cub everywhere."

" Yes, I am. Bernie and I are on our way back from Denver, say hello to Ed for me."

When refueling the Cub, I knew that I had to be careful. The gas hose nozzle was designed for large airplanes and dispensed fuel at a high rate of flow. In addition, the diameter of the nozzle was only marginally smaller than the Cub's tank opening. Cub pilots universally refer to these as DC-3 nozzles. Not being able to see into the tank, I over filled the Cub's tank and gasoline ran all over cowl and the hot exposed parts of the engine. I was a bit panicked by the thought of a fire, but the spilled gas simply evaporated.

Even after refueling at Dodge Center, it was necessary to stop again for fuel at La Crosse. It was only the second airport on the whole trip to have a control tower. The TKM radio has made communication with the towers a lot less problematic than it was in the past.

The leg from La Crosse to Waunakee was getting so familiar to both Bernie and me, that we really did not need maps. But after the debacle at Sterling, I kept one on top of the instrument panel just in case. The usual check points slowly passed underneath us. Once the Baraboo Bluffs were in sight it was only thirty minutes until we were back on the ground and had the Cub tucked in its hangar.

Postscript

These two trips were my sixth and seventh trips to Denver, After the first several trips, I felt as if I was wearing a groove across Iowa. Thus, several different modifications of the "southern route" were employed to add some variety to the trips. Once I flew to Denver just to help my brother John with a problem in the engine of a recently purchased Stinson 108. Flying to Denver is one of my favorite trips. Most of the time the wind is not a major factor. The weather is generally good, but sudden thunderstorms are an ever present concern. The more desolate the prairie, the greater the navigational challenge and the more "fun" I have.

Chapter 17
Out West With Bernie

As told in Chapter 8, my daughter, Ann, and I never reached our goal of Seattle, Washington. When I retired from the University, in January of 1998, I decided this would be the year to make another attempt. In Seattle, I would be able to visit Bob and Diane Dempster whom I had met at Lock Haven. There also was the possibility of attending the Northwest EAA Fly-In at Arlington, Washington. I discussed the trip with Bernie who was enthusiastic about it. A side trip to Eugene, Oregon was added to the itinerary, so we could visit Bernie's brother Vince and his family.

A CD ROM disk with information about all the airports in the western states aided in the planning of the seven map trip. My brother, John agreed to go along in a Super Cub and meet us in Rapid City. With all the plans in place, Monday, July 6, 1998 was set as the departure date.

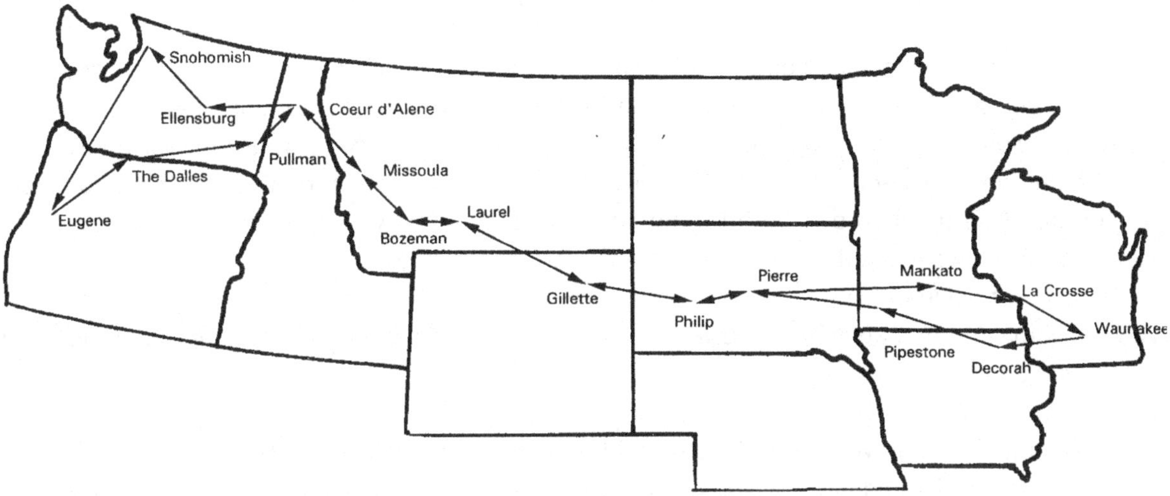

Route Map: Seattle Trip 1998

The Cub lifted off at 6:51 A.M. on the scheduled day. Bernie and I flew due west, under a grey overcast, towards Decorah, Iowa. In a half-hour, the Cub was flying just above the bluffs along the north side of the Wisconsin River valley. A mile or so past the Lone Rock Airport, the ceiling lowered and it began to rain. Ten minutes later, the visibility decreased to a half mile in heavy rain. Out the front, all I could see was rain on the windshield and the edges of the river bluffs popping out of the mist. I thought it best to return to Lone Rock, but Bernie said the Boscobel airport was only a few miles ahead. At an altitude of 200feet, the river was followed to the airport and we landed there. The weather showed no signs of improving, so Bernie and I had to wait. The only person at the airport was a retired Northwest Airlines pilot. He knew several pilots I used to fly radio control models with when I lived in Minneapolis.

Two hours later, the sun was shining through broken clouds, so it was back into the Cub and off to Decorah, Iowa. By the time Decorah was reached, the ceiling was down to 400 feet with a visibility of barely three miles. After refueling, we took off and headed toward Fairmont, Minnesota--134 miles away. A few miles past Cresco, Iowa, low clouds and heavy rain were encountered. I did a 180 and flew the Cub back to the

Cresco airport.

Now, we needed fuel, but the airport was unattended. I walked the few blocks into town and filled the two-gallon plastic gas can that was carried on this trip for just such circumstances. The 2 gallons of auto gas were poured into the Cub's tank. Taking off again at 1:00 P.M., Bernie and I headed for Fairmont across farmland planted in corn and soybeans. The north-south and east-west roads form squares that make navigation easy. One sets up a compass heading and then keeps the Cub's nose at the proper angle with respect to a road. Whereas, the typical general aviation pilot would be at 2,500feet(AGL) or higher, following the heading on their GPS display, and paying little attention to the details of the land below them. South of Austin, Minnesota, the ceiling went up and the visibility was at least ten miles.

After stopping at Fairmont for fuel, the Cub was flown over flat farmland towards Pipestone, Minnesota--96 miles away. Soon after we landed there, the deteriorating weather produced low white scud clouds that were skimming the tree tops and light rain. While waiting out the weather, Bernie and I spent the time chatting with the FBO and another fellow. Periodically, the Huron weather was checked on the computer weather system. At one point, the Huron weather was reported as a 400foot ceiling and ½ mile visibility. A half hour later they were reporting clear and 10 miles. Although it was still raining lightly at Pipestone, good weather was moving in our direction.

Strapping on the Cub again, Bernie and I took off in light rain and headed for Huron. After flying for about 20 minutes, the rain ended and blue skies could be seen to the west. Fifteen minutes later, the Cub flew out from under the cloud deck into clear skies. We arrived in Huron just after 5:30 P.M.. Our en route ground speed was 77 mph, which is very good for a Cub. We were concerned about getting gas as most fixed base operations seem to close up shop at 5:00 P.M.. Fortunately, this one stayed open until 7:00 P.M.. Although it now was after 6:00 P.M., we pushed on to Pierre arriving there at 7:44 P.M..

There was no trouble getting a room at the Best Western motel. I phoned John and learned that he had not reserved the club's Super Cub in time. He was going to fly his Bellanca 14-13. We agreed to meet in Rapid City at noon the next day.

The next morning there were blue skies without a single cloud. The night before, John had suggested that we fly on past Rapid City and now this looked like a really good idea. I called John and we agreed to meet in Gillette, Wyoming. He had to take Joye to the Denver airport so she could go to a 99's meeting in Mexico. As a result, he would not be taking off until sometime after 9:00 A.M.. In that Bernie and I had planned to spend a leisurely morning, we were now in a position of leaving Pierre in a hurry. When I received the next month's VISA card bill, I realized that I had forgotten to cancel the motel reservation in Rapid City. Cost me $100.

In the country west of Pierre, the navigator has to rely on creek outlines, roads and power lines for checkpoints. For some peculiar reason, the people who make the aviation maps will put a small dirt road on the map and ignore a blacktop highway. Again, Philip was scheduled for a gas stop. On this leg it was Bernie's turn in the front seat and once off the ground he flew the Cub. He always does a superb job of holding the assigned compass heading, altitude, and speed while I have some trouble staying on the course heading. Every time I look at something interesting, not on the course line, the nose of the Cub follows my gaze. In a few minutes, we are off course. Bernie is much more focused than I am and does not have this problem.

At Philip, an old Army surplus gasoline tanker was being used as a fuel depot—thanks to the EPA. It was being filled from a distributor's tank truck. The FBO said we could fill the Cub from the hose at the other end of the tanker. Being in a hurry, I said OK, but it was a decision I was to regret several days later. The fellow at Philip was restoring a Cub. Bernie and I had seen it before on our way back from the 1996 Denver trip. The rebuild was further along now, as the plane was assembled, and the engine installed.

Bernie and I decided to bypass Rapid City and go directly from Philip to Spearfish, South Dakota. The 111mile flight is over open country, but there are numerous checkpoints all along the course line. After what seemed like a long time, the Cub finally flew past the Sturgis airport. Bear Butte could be seen off to the northeast. The Spearfish airport is named for Clyde Ice who was the aviation pioneer in this part of the world. The lineboy did not know that Clyde had grown up in Waunakee, Wisconsin, but he did know some girls named Ice.

Now it was 1:00 P.M. and getting quite warm. Although we took off with no problem, the Cub was dragging its feet with respect to climbing. To get to a comfortable altitude, I had to circle the airport a few times before heading cross-country.

From Spearfish, the flight plan was to follow a broad valley around the north end of the Black Hills until it went off into the open country. What I had not planned on was that the valley had a number of high ridges on its left side. Once on course, the Cub started to have trouble climbing over them. Because of this, the Cub was flown to the northwest in order to follow Interstate 90. All went well until the Interstate started to go uphill at an alarming rate. The map showed the highway going up to 5,200feet as it went around a 5,829foot knob. The poor old Cub was struggling just to get up to 4,500feet. At a high density altitude, a rather high angle of attack is needed to fly level, the controls acquire a mushy feeling, and a couple hundred extra engine RPM's are needed. None of this gives the pilot a lot of confidence that altitude can be maintained. To improve the situation, I turned before the knob and went back down into the valley that now was lower than the Interstate. This maneuver reduced the problem and the Cub seemed to like the lower altitude. All of this was a bit disconcerting as I had flown the Cub through this area several times in the past with no difficulty. Bernie and I were paying the price for the 35 pounds of luggage stuffed above the baggage compartment. This was in addition to the tools, tie-down equipment, and extra bottles of oil that were stored in the compartment. The combined weighed of these items and the luggage was far beyond the 20 pounds allowed in the baggage compartment. The Cub's gross weight now was much higher than it had been on the earlier flights. It also was a lot hotter than it had been in the past.

Just before Gillette, there is a good sized lake. West of it there are a number of large buildings at the underground coal mines. The Gillette control tower, on top of a tall concrete structure, was a major improvement over the portable control tower on a truck of past years. Bernie set up the proper frequencies on the radio, I called in and had no trouble talking to the control tower. As the Cub entered the traffic pattern, John's red and white Bellanca could be seen parked at the fixed base operation.

The Gillette airport has a new airline terminal befitting a much larger city, but handles only four United Express flights a day. It now was 2:43 P.M., so the three of us took time for lunch at the airport restaurant. It was run by a young Vietnamese woman whose mother sat at a rear table observing us.

In that it was still a beautiful day with clear skies and little wind, we decided to press on as far as we could (my first rule of cross-country flying: if the weather is good--keep going). I really wanted to get to Laurel, Montana today. To do so would enable us to fly over Bozeman and Butte passes early in the morning when it was cool.

By the time we had lunch, refueled the Cub, and took off it was 4:04 P.M.. The line of flight to Sheridan took us over largely uninhabited country with a few gravel roads on top of the ridges and a railroad secreted in the valleys. The railroad made big sweeping turns as it worked its way around the seemingly never ending ridges. One of our checkpoints was a small town of Ulm alongside this rail line, but neither Bernie nor I saw it. I think it was a half mile or so off to the west, but we needed to fly right over the town in order to see it.

To the northwest of us was an area of rain in which lightning was very active. John's red and white airplane disappeared into the gloom near the thunderstorms. After a while, Bernie saw him flying south. I kept waiting for the Bellanca to come back north and fly by us. After what seemed to be an unusually long time, John finally gave us a buzz and headed north toward Sheridan. Then, the Bellanca slowly disappeared from sight.

Bernie and I were still having trouble finding good checkpoints and maintaining a decent clearance above the tops of the ridges. When we finally cleared the last ridge, Bernie's heading had us splitting the runway at Sheridan right at his ETA. It took us an hour and 21 minutes to fly the 77 miles from Gillette to Sheridan for a ground speed of 57 mph. The Cub's engine burned exactly 5 gallons of gas.

Due to construction, the main runway at Sheridan was closed. The remaining runway (23) was only 4,400feet long. The temperature was now in the mid 90's and the density altitude was quite high (we never asked how high). After the usual tasks were completed, I taxied the Cub out to the northeast end of runway 23. Checked for traffic and then pushed the throttle to full power. The Cub started to roll and as speed built up the runway was disappearing behind us at an alarming rate. Finally, at the 3/4 point, with a slight backwards tug on the control stick, the Cub lifted off the ground to an altitude of about 10 to 15 feet. That's all the higher it wanted to go. Upon reaching where 23 crossed the main runway, I made a shallow turn to the northwest and flew along the main runway that ended at the top of a tall cliff. Unfortunately, there was not enough of this runway remaining for the Cub to land on it. Far below the end of the runway was broad valley going off to the north. The Cub, Bernie and I went sailing off the end of the runway. Much to my surprise, the Cub started to climb rather than sink. We had it made.

Questionable takeoffs are not limited to slow climbing Cubs. When I finished combat crew training at Langley Field, Virginia, I was in a group selected to ferry Douglas B-26's from McCellan AFB, California to Japan. At the end of WWII brand new B-26's were flown from the factory to the storage facility at Kingman Arizona. During the Korean War, they were flown out of storage to McCellan and restored to like new condition. My regular pilot had grounded himself for fear of flying. He had not flown since WW II and could not get his skills up to what he felt were safe standards. I was assigned to fly with a ferry pilot, Bill Hanson, who had been serving as a flight test pilot at McCellan. In order to fly the B-26's from McCellan AFB to Honolulu, a 1,900 gallon rubber fuel tank was installed in the bomb bay. With this tank full, the B-26 was well above its normal gross weight. After waiting nearly a month for favorable winds, the group was finally given approval for departure.

At dawn, the six B-26's were lined up on the taxiway ready to go. When our turn came, we trundled out onto the runway. As we did , Bill ran his hands around the cockpit doing the pre-flight checks. Normally, the copilot reads the checklist to the pilot, but Bill had been flying B-26's every day for many months and felt that he knew the list. He stood on the brakes, revved up the engines to full RPM and released the brakes. The B-26 started to roll down the 5,000 foot runway. At the point where he was used to rotating, Bill pulled back on the control column, but nothing happened. The B-26 was now going well over the 140 MPH takeoff speed and was still on the ground. Just off the end of the runway was a river dike whose top was about ten feet higher than the runway. As the B-26 hurtled along the runway, Bill and I watched the dike get closer and closer. I started to think about what a funeral pyre the 1,900 gallons of high octane aviation gasoline in the bomb bay would make for us.

Finally, at about 200 yards from the end of the runway, Bill pushed full forward on the control column. The nose wheel oleo strut compressed. When the nose dipped, he hauled back on the control column. The compressed oleo strut gave the nose a strong upward thrust, the wing an angle of attack , and the B-26 lifted off the ground just above stall speed. The plane barely cleared the dike and went skimming across the river flats.

After a mile or two, the landing gear was retracted and the airspeed built up to the point where the B-26 would climb.

Once up to a thousand feet or so, Bill started to look around the cockpit to see why we could not get off the ground. It quickly became apparent that he had forgotten to lower the flaps to their takeoff position. After that, I read him the complete checklists each time we took-off or landed.

To the west of Sheridan, the Bighorn Mountains rose sharply. Our line of flight to Laurel, Montana took us on a northwest heading across the foothills of the mountain range. There were enough landmarks including the two power lines to keep us right on course. The Yellow Tail dam and the Fort Smith airport are located about half way to our goal of Laurel. A few miles past this point, north end of the Bighorn Mountains taper down and the Rocky Mountains can be seen off in the distance to the west. Unfortunately, the two power lines ended at the dam.

The countryside now consists of a series of long flat topped ridges, that are about 5 miles apart. There are high cliffs at their west ends and the ridges taper down toward their east ends many miles away. Flying along in the late daylight, each ridge seemed to rise up on the horizon as an unsurmountable height. But as the Cub approached them, they slowly seemed to lose altitude. Some would pass beneath on their own, while for others the Cub had to climb a few hundred hard-won feet to clear them.

There were some widely separated creeks flowing to the northeast, but they were of limited use as checkpoints. Bernie was unable to match them to those shown on the maps. Nearing the end of the leg, we could see the buildings in Billings, Montana eight miles to the northeast. Our attention now focused on the last high ridge before Laurel. Once over the ridge, it was clear that Bernie had us right on course to the airport that was only a mile or so to the northeast of Laurel. Bernie and I landed at 7:10 P.M.. John had landed a few minutes earlier.

The airport office was closed and there was no outside pay telephone. John went walking around the hangars to see if there was anybody at the airport. In one hangar, he found a fellow working on a Super Cub, who had a cell phone. He told John to call the Traveller's Motel. The man who owned it was the father of the guy running the fixed base operation. In due time, a grey Buick appeared. The driver took us to a motel that was straight out of the 1930's

After putting our luggage in the room, Bernie, John and I walked the eight blocks to the Owl Restaurant. Some fellow at one of the tables kept insisting to the waitress that he knew John. To which John replied: "If he works for the government, I don't know him". John and the fellow had a prolonged discussion relayed back and forth via the waitress. The antics finally ran out of steam and we had a very late dinner.

The next morning the three of us walked back to the same restaurant for breakfast. Around 7:30 A.M., the motel owner drove us back to the airport.

Since the leg to Big Timber was only 59 miles long, so it provided an opportunity for Bernie ride with John in the Bellanca. The ground elevation rose about a thousand feet between Laurel and Big Timber, so the Cub was back in the climb mode. Being alone in the Cub made climbing a snap and I flew over several high ridges with ease. The whole time, I could see Interstate 90 a few miles to the left of course. The airport at Big Timber was basically a cropduster operation. The Cub was refueled there, so we would have enough fuel for the climb up and over Bozeman Pass that was at 6,002feet.

With Bernie in the Cub's back seat, we took off from Big Timber, into clear skies and light winds. As the Cub struggled for altitude, John's Bellanca could be seen down in the valley buzzing farmers in the hay fields and making passes down the Livingston airport's runway.

Opposite the airport, the Cub was flown along the south side of a high butte where there was a small amount of lift. Finally, we worked the Cub up to the same altitude (6,312feet) as the top of the butte. With all this altitude, I headed for Bozeman Pass. En route another hundred feet were gained—the advantage of flying in the cool of the morning.

Interstate 90 went over the top of the pass and then downhill through a wide V formed by the edges of two mountains. On the south side of the V there were numerous vertical ribs of light grey rock running from top to bottom with pine trees separating them. This gave the cliff face a rather dramatic appearance. The Cub dove through the V and out into a large mountain basin. Bernie turned on the radio. The airport's UNICOM channel was filled with people reporting on a downwind leg, taking off, passing over the airport, and other things. I heard one pilot complain that someone had buzzed him while he sat on the end of the runway. What actually happened was that while John was on final approach, this guy taxied onto the runway forcing John to make a go-around. I worked our way into the traffic pattern and landed the Cub on the long concrete runway.

John had parked where there was a credit card gas pump. While refueling the Cub, a couple of young fellows drove up to look at the Cub. They were involved in the glider operations on the field and also flew a Cub. I asked one of them how to get enough altitude to go back east over the pass. He pointed out a mountainside that had lift when there was southwest wind. While on the ground, I told John that after getting over the pass at Butte, the plan was to fly on to Bowman Field just outside of Anaconda. There we would get fuel for the leg to Missoula.

Although the temperature was rising, Bernie and I had no trouble taking off from Bozeman and starting our climb. Halfway to the pass, there was a mountain whose south side was cut away from its base all the way to the top. I could not tell what they were mining, but huge trucks were dumping rock over a cliff that appeared to be 1,000feet straight down. It must have been exciting to back a truck up to the edge.

I had planned to go over Pipestone Pass (6,453feet) again, but the fellows at Bozeman had recommended we follow Interstate 90 over the higher Butte Pass. They indicated there usually was some lift from the sun hitting the rocks. From the east, the ground climbs gently up to the altitude of Butte Pass--the maps did not list its height. The Cub was climbing nicely and at the pass had reached an altitude of 6,800feet. Right at the pass, a downdraft cost us a few dozen feet of altitude, but not enough to interfere with getting over it. As the highway goes over the pass, it makes a sharp right turn. Then it makes a long run across the steep west face of the mountain.

Once past Butte Pass, the Cub was headed towards Bowman Field, near Anaconda, Montana, which has a credit card fuel system. As we flew over the Butte airport, Bernie saw John's Bellanca parked on the general aviation ramp. I circled for a while to draw his attention, but to no avail. I had to dissipate my hard won altitude to land. I did not shut off the engine for fear that it would not start in the 95 degree heat. I told John that he was at the wrong airport and that Bernie and I would meet him at Bowman Field.

I taxied the Cub to the south end of the 9,000foot runway and started to takeoff. Again, Bernie and I went roaring down the runway at a frantic speed before lifting off about 100 yards from the end of the runway. The Cub reached an altitude of 15 feet above the ground then refused to go an inch higher. I could see trees a city block ahead and realized that I could neither climb over them nor land in the rock strew field before them.

A flat turn was made to the right using mostly rudder. Then, the Cub dropped down into what I thought was an abandoned gravel pit with some brush in it (Bernie thought it was a swampy area). As the Cub flew through the depression, Bernie came on the intercom and said:

"I'm looking UP at the runway!"

Our immediate problem was getting out of the depression. I pushed full throttle and put just enough back pressure on the control stick to lift the nose a few inches. This gave the wing a slight angle of attack without losing any of the precious airspeed. The result was a few feet-per-minute rate of climb. Fortunately, the depression was long enough for me to maintain this semblance of a climb. Finally, the Cub had gained some 30 feet and was able to get out of the depression. At about 10 feet off the ground, the Cub went across the airport at 70 degrees to the runway (the thought of landing never entered my mind).

Upon reaching the other side of the airport, Bernie came on the intercom again:

"We just passed over a funeral ceremony in a cemetery".

I wasn't sure why he wanted me to have that piece of information. Once over the grass, the Cub decided that it had given us enough of a thrill for the day and started to climb reasonably well.

The Cub was circled over the edge of the Butte airport until enough altitude was gained to leave the area. As is always the case, there was a high ridge to climb over in order to get to Bowman Field. The Cub was equal to the task and the airport was reached at 1:22 P.M.. John was waiting there for us. Bernie and I told John about the takeoff at Butte, but he had been taxiing away from us and had not seen the event.

Landing at Bowman Field cut the Missoula leg to 99 miles. A more comfortable distance than the 124 miles from Butte. While we were refueling the Cub, a small forest fire could be seen halfway up the mountains to the east. It had been started by lightning during the night.

Bowman's runway was 6,000feet long and the temperature was in the mid 90's. However, the Cub got off the ground smartly. Bernie and I proceeded to climb the Cub on course to Missoula without the usual preliminary circling to gain some altitude. This segment is mostly a sight seeing leg as it goes through a mountain valley that has many signs of civilization: ranches, small towns, etc. The valley varies in width from broad to narrow. It is very interesting to fly along, peer up the side valleys, and look up at where the mountain tops had been clear cut.

We simply followed the Interstate highway while staying high enough clear the numerous power lines that cross the valley floor. When Ann and I flew this same valley in 1978, the power lines had large red balls on the top wires to identify them. This time there were no balls. Bernie and I had to continually scan the mountain sides for power line towers. On the old sectional maps, this valley was designated as a low altitude route through the mountains and was identified by blue diamonds on the maps. The new sectional maps do not show the route.

John would occasionally fly past us at a higher altitude, circle around behind us, then make a high speed pass underneath us. Thirty miles from Missoula, a large side canyon had a huge set of power lines going across it that were supported by tall white and red towers. Strung across the mouth of the canyon was a diagonal cable with a number of strobe lights blinking rapidly. The power lines and cable completely blocked the entrance to the canyon. At this point, the main valley made a ninety degree turn to the right.

Bernie and I were contentedly flying up the valley, when suddenly it was blocked by a ridge that the Interstate highway zig-zagged through. The city of Missoula was on the other side. Bernie set up the radio frequencies. Since John's handheld radio had quit working, I called in as a flight of two. The tower asked me for the N number of the non-radio airplane, but I did not know the number. John came alongside of us with his landing gear and flaps down and was flying at Cub speed. The tower cleared us for a left downwind to runway 11. I set up for what I thought was the runway, but it had a big 7 on it. I looked right at the number, but it did not register as the wrong runway. Finally, when I started to cross runway 11, the tower called up and told me I was set up for the wrong runway. I made a quick heading correction and entered the traffic pattern for runway 11. After landing, the Cub was taxied to the ramp and tied down. The Cub's ground speed on this leg was a less than sterling 54 mph.

By this time the three of us were very tired, so we decided to RON here. The infamous Mullan Pass would be tackled in the cool of the morning. We were looking over the motel selections when the gas truck pulled up to our airplanes. As John and I dashed out the door, I yelled to Bernie "find a motel that will pick us up." I filled up the Cub's tank and my two-gallon plastic gas can. When John and I came back, Bernie had reserved a room at the Doubletree Inn that was the most expensive motel in Missoula. Actually, we were fortunate to get the room as there was a national convention of BMW motorcycle fans. Rooms were in very short supply. For the rest of the trip the watch word was "don't let Bernie choose the motel."

For dinner we went to Fitzgerald's family restaurant that was built over the top of a ravine. There was a good sized creek at the bottom of the ravine. Bernie ordered a top sirloin steak, but it turned out to be simply a piece of meat cut from a pot roast, so much for family restaurants. After dinner, Bernie and I went for a walk through the eastside of Missoula. The bars were full of BMW bikers. Bernie talked to a couple about their motorcycles. I told them that this convention was like a high class Sturgis.

The next morning (Thursday), John decided to return to Denver rather than continue on the trip. We taxied out as a flight of two airplanes. After takeoff, John flew by, gave us a waggle of the wings, and the Bellanca disappeared into the valley toward Butte. The Cub, Bernie, and I headed toward the infamous Mullan pass that Ann and I had never reached. The plan was to get over the pass and then land at Shoshone County Airport, at Kellogg, Idaho. There, I would pour the two gallons of gas into the Cub's tank. This was necessitated by the 145 mile leg, from Missoula to Coeur D'Alene, Idaho, being at the limit of the Cub's range.

The valley has a fair amount of civilization along the Interstate highway. At about a third of the way, there was an airport called Mineral County. The paved runway was 3,400feet long with trees at each end. It looked to me that I could land there in a pinch, but getting out would be "interesting".

The pass that the Interstate highway actually went over was called Lookout Pass. Often, it is erroneously referred to as Mullan Pass. Its altitude was shown on the map as 4,738feet. By now I knew this was the altitude of the highway and several hundred feet more altitude would be needed. A slow climb to 5,000feet was initiated. As the Cub approached the pass, the valley became narrower, the Interstate highway went into the woods, and started going uphill. Off to the right was a mountain with a VOR building on its top. To the east of it was the real Mullan pass at an altitude of 5,260feet. However, we were not going to negotiate that pass. It is worth mentioning that it was another clear day with no wind and unlimited visibility.

By the time we reached Lookout Pass, the Cub was about 200 feet above the highway. Bernie and I could read the license plate numbers on the cars parked in the scenic overlook. The Cub had plenty of clearance and we flew over the pass in good style.

In 2001, at Hartford, Wisconsin, I talked to a fellow who flew a Piper PA-12 over Mullan Pass in bad weather and high winds. Just as he went over the pass, a downdraft caught him and the left wing hit a tree top. The wing fabric was ripped and the aileron bent. He was able to regain control and flew to an airport where temporary repairs were made.

On the west side of the pass, it was only 25 miles to Kellogg, Idaho, so I started a gentle letdown. At the town of Wallace, the Interstate seemed to disappear as it went into the town. The Cub passed over several more small towns hanging on the sides of the narrow mountain valley. By the time we got to Kellogg, the gas gauge wire showed there was quite a bit of gas in the Cub's tank, so we kept on flying. At Rose Lake, the Interstate highway disappeared into a series of ridges that were higher than we were flying.

About this time, I decided we did not have enough fuel to get to Coeur D'Alene. I turned the Cub back toward Kellogg. Bernie came on the intercom and told me that he thought we had enough fuel to reach Coeur D'Alene. After a short discussion, Bernie convinced me. I turned the Cub around again and headed toward Coeur D'Alene. Unfortunately, I also had to climb over several ridges to get there. Fifteen miles from Coeur D'Alene, I could see the broad valley where it lay. I cut around the peak of a high hill and headed the Cub for the airport.

When the Cub was parked, the two gallons in the plastic gas can were poured into the Cub's tank and the gas truck added another 6.8 gallons. At a ground speed of 64 Mph, we had made the 145 miles in two hours and ten minutes on only 8.8 gallons. This was a near record distance for me. One year on a trip to Lock Haven, I had flown 140 miles on two different flight segments in dead calm. Once I had a tail wind and flew the 156 miles from New Castle to Lock Haven nonstop. A record not to be broken easily.

The last time that Bernie and I had flown to Colorado most of the trip was in 95-100 degree weather. On that trip, I had been using 80 weight oil in the engine and was getting only six hours per quart of oil. Because of this, 100 weight oil had been put in the engine for the present trip. Now, the engine had barely used a quart in 25 hours of flying time in the 95 degree heat--a major improvement. I rounded up an oil bucket, drained the old oil, and put in four new quarts of100 weight oil.

Takeoff from Coeur D'Alene, under clear blue skies, was at 11:55 A.M.. Lunch would be missed yet another day. The route of flight, 125 miles in length, was to a cropduster strip at Davenport, Washington then on to Ephrata, Washington.

As Bernie and I headed out on course, there was a high ridge right in front of us that would have required a lot of climbing to get over. To avoid the ridge the course line was shaded a bit. I soon found myself looking at the end of the runway at Felts Field that had a control tower and a 5-mile control zone. The airport on the ground was not where it seemed to be on the map. The Cub was turned to due north and we ducked around the end of the ridge. I hoped that the tower operator, who was busy with an airplane on final approach, did not see the Cub. On the back side of the ridge, there was a Cessna 150, at our altitude, heading the other direction. Hopefully, he would distract the controller.

On the way past Spokane, the municipal airport and Fairchild Air Force Base could be seen. The countryside became flat and dry looking although there were many circles of irrigated wheat fields. As our ground speed was better than expected, Bernie and I overflew the cropduster strip at Davenport and headed southwest toward Ephrata. The wheat fields were left behind and now we were flying over a light brown colored desert. At about the midpoint of this leg, there were a large number of pothole lakes. Some of which were on the map and many that were not. Thus, they were not much help to us. In that we were flying over a desert, I did not understand how there could be so many of these lakes. Strangely, not much of anything seemed to be growing around the lakes. Perhaps the water was toxic.

After a long time without any good checkpoints, a railroad track was crossed. Off to the northwest, we could see a small town. After flying to it, Bernie determined that the six houses and a grain elevator comprised the hamlet of Irby, Washington. It was 8 miles off our course line—so much for the 100-yard rule. We followed the railroad track northwest to the town of Wilson Creek, which was on our course line. It was slightly larger than Irby and had a small unattended airport. A new compass heading was established that would keep us on course.

As the Cub approached Ephrata, the Moses Lake airport was about 6-7 miles to our left. I looked out the right side of the Cub's cockpit, much to my surprise, I saw the nose of a huge Boeing 747 pointed right at us, but it was higher than we were. The 747 was on final approach to the Moses Lake airport. Later, I learned that Japan Airline was training their 747 flight crews at the ex-Strategic Air Command base.

While I was scanning the horizon for signs of the Ephrata airport, Bernie asked:

" Are those buildings to our right at the airport?"

" No, those are farm buildings."

After a few more minutes, I still could not see anything that looked like an airport and asked Bernie:

"Can you see the airport?"

He replied:"Look straight down."

The Cub was right over the airport. All I had to do was to make a 360 degree overhead approach and land. These approaches are a lot of fun as I can pretend that the Cub is a fighter plane returning from a mission. To accomplish this approach, the engine is set at low RPM's directly over the landing spot. A left hand turn with a constant rate-of-turn and rate-of-descent is established. All the while, I keep my eye on the landing spot and make subtle adjustments in the rate-of-turn and rate-of-descent to maintain a proper spiral. Just as the 360 degree turn is completed, the engine is idled, the wings leveled, and Cub's wheels touch down at the landing spot.

As we taxied in, there were glider operations going on. I tried to taxi around them, but the ground crews waved us back onto the taxiway. After parking the Cub, we walked into a hangar full of Russian YAK airplanes. Bernie asked about getting gas and was told that it was handled at the cropduster office next door.

After refueling the Cub with 8.4 gallons, the lady at the office let us take the courtesy car. It had "Port of Ephrata" painted on its side, which seemed out of context. I drove the car into town where Bernie and I had lunch at McDonalds. On the way back, a quart of Marvel Mystery Oil was purchased to help keep the engine in shape. Bernie and I were mystified as to how a town in the desert could be a port. The lady at the airport office explained that the Columbia River, at the edge of town, was large enough for cargo barges, hence the port designation. I assumed that most of the cargo was outbound wheat.

Taking off in 95 degree temperatures to begin the 56 mile leg to Ellensburg, Washington posed no problem. The Columbia River was followed south to where Interstate 90 went over the river on a high bridge. We were surprised at how wide the river was this far from the Pacific Ocean.

To the west of the river, the Interstate highway went up a long slope toward Ellensburg. The whole dun colored countryside had the appearance of being one smooth barren rock that went uphill en mass. The map completely deceived us as to what we had to fly over. The top of the hill was much higher than expected from the map. Climbing the Cub was tenuous as the rock was radiating the sun's considerable heat back into the already hot air. Each time I tried to gain altitude, the Cub did not seem to want to gain more than a few feet. In addition, our 54 mph ground speed was nothing to cheer about. I was beginning to have serious doubts as to whether the Cub would climb enough keep us above the hillside. After what seemed a very long time, the top of the hill was reached, and a mountain basin was entered. In another fifteen minutes, the Cub was landed at the Ellensburg airport.

When Bernie and I were ready to depart, the Cub's engine decided not to start. After fiddling with it for a while, a fellow flying a VARIEZE came over to give us a hand. He was waiting to meet a buddy of his who also flew a VARIEZE. Then, the both of them were flying to the Arlington, Washington EAA Fly-In. After propping the Cub for a long time, the engine finally sprang to life with a roar.

A few miles after takeoff, the valley leading to Snoqualmie Pass was entered. There were four sets of power lines strung at different heights across the valley's entrance. They looked like a huge grey curtain blocking entrance to the valley. I don't recall seeing any red balls on the top wires. Fortunately, we were high enough to clear them, but they could have been a real surprise in bad weather.

Initially, the valley was reasonably wide, but as we progressed along our line of flight it became much narrower. Eventually, even the bottom of the valley was covered with pine trees. A few miles before the pass, there were three large lakes (formed by dams on rivers) on the right-hand side of the valley. The pass was a few miles past the lower end of the third lake.

Since the pass was shown on the map as being at 3,002feet, I had climbed the Cub up to 3,500feet. I thought that the Interstate would go through a valley to the right of the mountain directly ahead of us. As we reached the pass, a surprise was in store for Bernie and me. The pass was at the bottom of a several thousand foot tall mountain wall. To the left of the wall was a high ridge that slanted down to meet it. At the T junction, the Interstate highway made a 180 degree turn around the end of the ridge. The slot formed by the end of the ridge and the mountain wall was too narrow to fly through. Seeing this, I flew the Cub back to the side of a mountain a mile or so before the pass. Then, I proceeded to work the Cub up to a much higher altitude. After five to 10 minutes, the Cub was at a high enough altitude for us to see over the ridge and into the tree covered valley beyond it. Going over the ridge Bernie and I looked down on a ski area where Ann had skied, when she lived in Seattle.

Once over the pass, we flew twenty miles down an ever widening tree covered valley. There was a high overcast at the end of the valley, but the visibility remained excellent. Just as the end of the valley was reached, the two VARIEZE's went past us and disappeared off to the northwest.

Flying toward Snohomish and its Harvey airport, Bernie and I got our first lesson about navigating in the Pacific Northwest. The map would show a town, but all that could be seen on the ground was an amorphous cluster of houses hidden in the trees with no city center and worst of all, no water tower. I never did figure out why the towns here do not have water towers. We muddled our way along the course line. As usual, Bernie's navigation brought us right to the small airport. The 100 miles took two hours to fly, at a ground speed of only 50 mph, and the Cub burned eight gallons of gasoline.

The airport had a grass runway, quite a few hangars, and a restaurant. On its patio, people were eating meals and watching the activities. The Cub was tied down in a nice grassy area. Bernie and I had dinner in the excellent airport restaurant (the first decent meal we had in a long time). We reserved a room at the Snohomish Inn. Bernie elicited several different opinions as to how far it was to the Snohomish Inn. Finally, he and I decided to walk there, but it was a lot farther than we thought.

Friday morning, Bernie and I arose at 6:30 A.M.. The cab was there at 7:20 A.M. to take us to the airport where we had an excellent breakfast.

To fly the 18 miles to Arlington, the instructions were to fly over the Green Valley airport and report in with the Fly-in's approach control. This airport was a grass strip nestled in the middle of a wooded area. I scanned the area, but could not see the airport. Then, I called in on the radio, and talked to a woman on the ground. I told her that I was about a quarter mile from a column of smoke, she said that I should be right over the airport. Bernie tapped me on the shoulder and told me to look straight down, sure enough it was right below us. The woman told me to follow a GLASSAIR to Arlington. The plane whizzed past us, at about 200 mph, and immediately disappeared from sight. I made a few remarks to her about the futility of a Cub following a GLASSAIR.

In a few minutes, we could see the Arlington airport. I proceeded to get the Cub into the traffic pattern and then landed. The ground control directed us to the classic area where the Cub was parked next to Bob and Diane Dempster's Super Cub. A note was left on their Cub's front seat telling them we were parked next to them.

Bernie and I had made arrangements to meet a Mankato High School classmate of ours, Bill Hawes, who arrived at 10:30 A.M. sharp. Bill said that he had a pick-up truck at home that Bernie and I could use for transportation. We drove to his home in Kirkland that was in an area of large lots. Bernie and I had not seen his wife, Nancy, since skiing with her and Bill, in 1950, at Hayward, Wisconsin. Bernie and I returned to the Fly-in via Bill's white Dodge truck. The Cub was left overnight at the Fly-in.

That evening, Bernie and I took Bill and Nancy to dinner at an Outback restaurant (my turn to buy). Later, while watching TV, we learned that, in the afternoon, Ephrata had severe thunderstorms with 60 mph winds and golf ball sized hail. Fortunately, we had gone through Ephrata the day before.

The next morning (Saturday) Bernie and I drove back to the Fly-in and spent the day wandering around. I thought the Fly-in was rather poorly attended with respect to antique airplanes. Some of the locals said that the owners of the antique airplanes and the EAA group sponsoring the Fly-in had developed some hard feelings. The former had stopped coming to the Fly-in. I was very disappointed as this part of the U.S. is known for its antique airplanes. War birds were almost nonexistent. The best one was a homebuilt 2/3 size Spitfire.

Back at the Cub, I saw Bob Dempster walking to his airplane. I ducked around my plane and sat on the ground facing his airplane. He read the note then looked to the left and then to the right. I waved at him and said "Hi Bob." He was surprised to see me, although he knew we were planning to be at the Fly-in. Bernie and I talked with Bob and Diane for quite a while. They were living in a condo on the edge of the Renton airport. Diane could walk to work at the Boeing plant.

They brought Bernie and me up to date on their trip around the world in their Super Cub. En route from Iceland to Ireland they had a complete electrical failure and had to depend upon their handheld radio and GPS. Diane made radio contact with a transatlantic Delta Airlines flight and told them:

"This is Super Cub N4155R, we are 350 miles out of Reykjavik en route to Shannon, Ireland."

The incredulous Delta pilot replied:

"You're what and you are where?"

Diane did not relate the rest of the conversation to us.

After leaving their Super Cub in England for a winter, they started out on a flight to Australia. All went well until they reached the western borders of Thailand. An air traffic controller came on the radio and said:

" Super Cub N4155R climb immediately to ten thousand feet".

A startled Bob, immediately followed instructions, climbed to 10,000 feet, and leveled off. Once there he called the controller:

"Why the sudden change in assigned altitude?"

He replied: "Insurgents in the jungle are shooting at airplanes."

The rest of the trip to western Australia was uneventful. After spending some time flying to various cities to visit friends, they left the Cub in a hangar for another winter.

The next summer, Bob and Diane flew the Super Cub to Japan. They had tried to make arrangements to fly to Russian Siberia and refuel the Cub there before the last leg to Alaska. However, their Russian contact failed to make the proper arrangements and the rest of the trip was canceled. Bob and Diane, flew the Super Cub back to Okinawa where they disassembled it, put it in a container and shipped it home. They spent a couple of years working out the logistics of flying the Japan to Alaska route. They flew the Super Cub the 1,700 miles from Seattle to Oshkosh in 16 hours non-stop to prove that the flight from Japan to Alaska could be made. Things did not work out and the last leg of the round-the-world flight did not get flown. Eventually, all the extra equipment was removed and the Super Cub was returned to its original configuration, but plastered with decals from the countries they had visited.

Late in the day, the Super Cub, that John was supposed to have flown, was right behind me as I taxied to the runway. I flew back to the Snohomish airport and then tied the Cub down for the night. Bernie drove the truck and met me there. On Bill's recommendation, the four of us went to an excellent seafood place for dinner (Bernie's turn to buy).

Sunday morning, Bill took us to Seattle where we visited the locks between Puget Sound and Lake Washington. The locks had a fish ladder. It was very interesting to watch the large salmon swim up stream and then leap over the steps of the fish ladder. The three of us ate lunch at the Museum of Flight and then waited for Bob Dempster to arrive. When he did, the four of us took a leisurely stroll through the museum with Bob providing considerable local color commentary. The best part of the museum was the old lumber mill building that was the original Boeing plant. In one corner was the restored office of Clair Egtvedt, who was Boeing's chief designer in the 1930's and eventually company president. On the wall was a certificate indicating he was a member of the Stoughton, Wisconsin Hall of Fame. John Vorndran had told me that Clair Egtvedt was a Stoughton boy, so I took some photos of the office and its plaques for John.

Monday morning it was time to go to Eugene, Oregon and visit Bernie's brother. Bill drove us back up to the Snohomish airport. We said our goodbyes, took off, and headed south towards the little town of Enumclaw, Washington. The temporary FAA approved route for flying through the Seattle mode C veil during the Fly-in was followed (unfortunately it had expired the day before). Much to our dismay, this leg offered some of the toughest navigation encountered on the whole trip. Underneath us it was mostly woods with scattered housing hidden in the trees. If a problem occurred, there were very few places in which to land the Cub .

After 30 minutes, the Cub flew past an airport, in the midst of the woods. Bernie identified it as Crest. It was six miles to the west of our course line. The Cub was put on Bernie's new heading for Enumclaw. After a while, we came to what I thought might be the town. Again, there was an amorphous collection of houses, no water tower, and the roads did not match the map. Despite this, Bernie gave me a new compass heading. Some time was devoted to climbing the Cub up and over a tall ridge that had a lake behind it. Finally, we saw a large cement plant off in the distance that was depicted on the map. The town of Eatonville, which was another of our turning points, never was seen by either of us. After an hour and 45 minutes, we landed at the Chehalis, Oregon airport.

The Cub's gas tank was filled and again the engine refused to start. Finally a mechanic came out to look at the engine. He removed all but one of the spark plug wires and checked the spark from the magnetos. The spark seemed to be OK. Then, Bernie and I proceeded to try at least a dozen different schemes to get the engine started. Finally after two hours, one of the fellows helping us got the engine started. I went in to pay them for their efforts, but they said no charge. I thanked them and ran back to the plane.

It was 103 miles, on a southerly heading, from Chehalis to McMinniville, Oregon. The first half of the trip was over tree-covered mountains. There were many areas that had been clear cut and reforested. We were flying low enough that we could easily see the six inch tall seedlings. As usual, Bernie and I expended considerable time and effort getting over the mountain ridges. Although they were not all that high above sea level, their tops were about 1,500feet above the lower terrain. As we flew past the Portland metropolitan area, the Cub crossed the wide Columbia River. After an hour and 45 minutes, McMinniville was reached. Howard Hughes' Spruce Goose resides in this city, but we did not see its home.

We refueled the Cub and again the engine would not start. It was the same problem. The engine acted as if it was vapor locked and not drawing fuel. Our whole repertoire of starting procedures was tried to no avail.

One of the local mechanics came by, so we gave him a run down of the events. He suggested disassembling the gascolator and checking the fuel flow. In addition, he detected some junk when he drained the gascolator. Bernie got a sheet of paper towel. The cowl was dropped and fuel drained out of the gascolator onto the paper towel. A circle, about 5 inches in diameter, of what appeared to be mud, was deposited on the towel. It was a miracle that the engine even ran with that much crud in the fuel system.

Bernie and I took the gascolator apart. There was more mud in the bottom part, but the fine mesh screen at the top appeared to be clean. The mechanic came by again, looked at the screen and said that it looked clean to him. Bernie and I concluded that the gas stop at Philip, South Dakota was the cause of the muck. When the distributor pumped gas into the tanker the turbulence caused the muck at the bottom of the tank to get stirred up. A mixture of gas and muck was pumped into the Cub's tank. Usually, the tanker's filters should stop this from happening. I suspect that the tanker either didn't have filters or they were already full of dirt. I should have known better as after the fuel storage tank at the Waunakee airport is filled, there is a similar problem with water in the fuel. Why the muck did not kill the engine, while we were in flight, thankfully remains a mystery. Once the gascolator was cleaned and reassembled, the engine started on the first flip of the propeller.

The leg to Eugene, Oregon was only 75 miles long. The line of flight was south along the Willamette River. In the flat farmland that filled the broad valley, the major crop was grass seed. Further ahead in the valley there were three large columns of white smoke that joined into one big cloud at the 1,000foot level. The light wind then blew the smoke east across the valley until it hung on the foothills. Later, Bernie and I learned that the smoke was caused by burning the fields to kill the weed seeds. At one time all the fields were burned in this manner. Now, only a few farmers are allowed to do so, but under strict EPA rules.

When we were at McMinniville, a fellow flying a Citabria gave Bernie and I a song and dance about how hard it was to land at Eugene because of its control tower. He recommended that we land the Hobby airport seven miles south of Eugene. Bernie was not too happy about this idea as Vince lived right under the Eugene traffic pattern. I told Bernie that we would go into Eugene if the radio worked (we sometimes had a bit of trouble getting the radio to turn on). Approaching Eugene, I called in as a no transponder airplane, did an identification turn, and proceeded to land. It was just like going into Madison before they got the ARSA. Vince had been listening to the tower on a radio at home. He appeared at the airport a few minutes after we landed. Bernie and I had flown 2,064 miles in 34 hours and 47 minutes of flying time for an average ground speed of 59 MPH. It now was Monday, July 13,1998.

On Tuesday, Vince took us to an auto parts store that had the right sized belt for the Cub's alternator. The old one had become very loose. It appeared to be slipping although the battery was staying charged. Back at the airport, I changed the belt, which involved removing the propeller. I also installed a small thermometer in a shady spot on the upper left side of the cockpit. Then, we visited the airport's Oregon Air and Space Museum where Vince was a volunteer. It had a small collection of aircraft and some memorabilia.

On Thursday, Vince drove us across the costal range of mountains to visit an air museum at Tillamook, Oregon that was housed in a WW II blimp hangar. The building was 120 ft high and as long as three football fields. It was constructed, in only thirty days, entirely out of bolted together wood beams. The museum had a good collection of WW II planes including a rare P-38, a Douglas B-26, and a SBD like those Vince had flown in the Navy during WW II. This was the place where the well-known aviation writer, Jeff Ethell, was killed flying a P-38.

We went over to a maintenance hangar where they had a beautiful Grumman J2F "Duck" biplane. Bernie was talking to the chief pilot about flying the warbirds who said that it was a bit stressful due to the age of the planes and their value. Recently they had a Corsair crash on the way back from an air show. The pilot bailed out at a low altitude, but was not hurt. When the Corsair hit the ground, only the wing tips could be seen outside of the fireball. The chief pilot asked Bernie:

"What are you fellows doing out here?"

Bernie replied: "We have just flown from Wisconsin in a Piper J-3."

"I envy you fellows, that is real flying".

Needless to say, both Bernie and I were walking on air for quite a while.

By Friday, both Bernie and I were getting to feel that we were overstaying our visit. Recall the old saying "fish and relatives get old in three days". Our original plan was to stay through the weekend in order to see an air show associated with the Oregon Air and Space Museum. Bernie and I concurred that it was not enough of a draw to keep us there an additional three days. Departure was scheduled for Saturday morning.

The Return Trip

The initial part of the return flight went from Eugene to the Columbia River Gorge. Then, the river was to be followed east to the city of The Dalles. From there the course went due east to Pasco, Washington and on to Pullman, Washington. Then, the course line went north to Coeur D'Alene, Idaho where it rejoined the route taken on the way to Seattle.

Saturday morning, we awoke early and were greeted by a low ceiling and limited visibility. Vince took us to the airport, then left to work at the air show. About 9:00 A.M., the ceiling was above 1,000feet with the sun peeking through the overcast here and there. Bernie and I climbed into the Cub, contacted the tower, and took off. The Cub was on its way to Aurora State Airport, just southeast of Portland. On the way there, a couple of airplanes were seen heading for Eugene. The Aurora State airport was quite busy, so it took a long time to get fuel.

The ceiling was at 1,500feet with good visibility under it, so after refueling the Cub at Aurora State, we flew northeast to the Columbia River gorge. As the Cub approached the gorge, the tops of the cliffs were still in the clouds, so the Cub was flown a few hundred feet below the cloud base and close to the south side cliffs. It was a bit disconcerting to look out under the right wing and just see the face of the cliffs. Proceeding east up the gorge, it became clear that there was no place to land. An Interstate highway, whose east and west bound lanes were separated by a fence, clung to the south side of the valley between the cliffs and the water. Needless-to-say, this resulted in a very high pucker factor.

Looking down, Bernie and I could see two airplanes, an Aeronca Chief and a Cessna 150, flying our same direction, but just above the water. The wind was going up the river to the east. Due to differences in the wind speeds at our two altitudes, The Cub kept pace with them for most of the length of the gorge. By the time we were half way through the gorge, the clouds started breaking up. The Bonneville dam was flown over. It appeared not to be very tall and was at a relatively wide point of the river. It is a "pass through" dam whose only purpose is to generate electricity.

Near the town of Hood River, there were a large number of windsurfers leaving white wakes on the river. There were so many of them that the surface of the river looked like it does when a school of carp thrashes around trying to escape from shallow water. The Hood River airport was high up a broad valley to our right. Even though it was reputed to have many antique airplanes, Bernie and I flew on past it.

At this point, the clouds were gone, the trees on both sides of the river were getting sparse, and the bare ground was becoming light brown in color. In addition, the hills on both sides of the river were getting lower and tapering off to the east. To the north, the snow-capped peak of Mount Adams could be seen. I was surprised at the short distance (45 miles) that it took for the terrain to change from lush pine forests to desert.

The city of The Dalles was finally reached. I set up to land to the west into the 10-15 mph wind. Then, I made the worst landing of my life. I was making a wheel landing, just about the time the wheels were to touch, the bottom fell out. The Cub hit hard and bounced into a half-stalled attitude. I added some power, the Cub went up about four feet into the air and then dropped back onto the runway. On the fourth bounce, I added a lot of power while pulling full back on the stick. The Cub decided it had enough fun and plopped down on the runway for good.

Luckily we did not suffer the fate of a fellow in a beautiful WW I Sopwith Camel replica I saw at a Fly-in. The plane came down on final approach with a high rate of sink. When the airplane hit the runway, it bounced quite high. It came down and bounced again. On the third bounce, the pilot gave full power. As the nose went up, the airplane torque rolled over on its back and it hit the ground inverted. The nose and the top wing took the bulk of the impact and the pilot was not hurt. Fortunately, the Cub kept its nose pointing straight ahead with the wings level during all the bounces.

At the gas pump, Bernie and I talked to the two fellows who had flown below us for most of the Columbia River Gorge. They did not seem to be a bit concerned about flying just above the massive sets of power lines crossing the river. They were planning to land next at a short forest service air strip in the mountains south of The Dalles. This did not seem to be the best of plans to me.

Our next stop was the Tri-Cities airport at Pasco, Washington--111 miles away on a 56 degree heading. We followed the Columbia River, now flowing through very dry looking, brown countryside. Initially, there was some high ground near the river, but as we went to the east the countryside became a broad relatively flat plain with no visible signs of vegetation. Forty five miles from Pasco, the Interstate took a bend to the southeast and faded from view. This happened at the same place on the map where there was a large restricted area—the Boardman Bombing Range

Fifteen miles from Pasco, the Interstate reappeared, now going due north. Bernie's compass heading took us to where the highway crossed the Columbia River. The Interstate was followed north up over a ridge. The Tri-Cities airport had a control tower, so Bernie set up the radio frequencies. I called in from five miles out and we landed. The ground speed for this leg was a nifty 76 mph. At the fixed base operation, the thermometer read 100 degrees. Up in the air, my new thermometer had been showing slightly more than 90 degrees.

After refueling the Cub and checking it over, we taxied out to takeoff for Pullman, Washington--107 miles away. It was calm, so I requested the longest runway. The tower agreed and I taxied to runway 23. On the takeoff roll, the Cub was using up the runway at a frantic rate, but it lifted off at about the halfway point. After turning away from the runway, I had no problem climbing the Cub up to our cruising altitude of 500feet AGL. I suspect the underlying reason for the good rate of climb was that the airport was only 407feet above sea level.

East of Pasco, irrigation booms provided water. The ground was literally covered with circles of green or gold wheat fields. The Snake River was crossed. It was our last good checkpoint for a long time. Off to our right, a series of grain elevators were located along an abandoned railroad track. As we could not tell which one was which, they were not a lot of help. A town called Clyde was on our course line, but we never did see it. Coming close to the Snake River again, Bernie finally saw several good checkpoints. He determined that we were only a few miles off course. By this point, we had flown 80 miles and had seen some crews combining wheat fields, but never a ranch or farm house. Except where it was irrigated, the land was light tan in color. It did not look good enough to graze cattle and we never saw any.

Approaching Pullman, Bernie and I could see the buildings of Washington State University. A few miles to the east, the buildings of the University of Idaho at Moscow also could be seen. The airport was halfway between the two towns. On final approach to runway 23, it seemed as if we were just skimming the tops of the hills. The ramp was on ground that was quite a bit higher than the runways. From there we could look down at airplanes taking off. The Cub was refueled, some oil added, and then we proceeded to start the engine. Nothing happened, Bernie and I went through our usual repertoire of tricks, but the engine refused to start.

After over an hour of propping the engine in near 100 degree temperatures, Bernie and I gave up. The lineboy reserved a room for us at the Holiday Inn Express Motel. They sent their van out to pick us up. Although there was Denny's across the parking lot, the desk clerk suggested Pete's restaurant a couple of bocks away. We opted for Petes. It was a combination bar and restaurant, but there were only two sets of people in the rather large dining room--a bad sign. Another marginal meal was had.

The next morning(Sunday), it was 6:00 A.M. before we were on the way to the airport. This caused me some concern as I wanted to get to Coeur D'Alene, Idaho early in the morning, so we could go over Lookout Pass in the cool air. Getting off the ground at 6:38 A.M., the Cub was headed north on a compass heading of 352 degrees.

The countryside along the route of flight was becoming a bit more mountainous with a 3,200foot peak here and there. Near the end of the leg we flew along the heavily wooded west side of Lake Coeur D'Alene. There were very few houses along the lakeshore. It would have been a very nice place to explore on the ground. It took us only 57 minutes to fly the 73 miles from Pullman to Coeur D'Alene--a welcome ground speed of 77mph.

When the Cub was topped off, I also filled the little two-gallon plastic gas can. It was placed on the floor under Bernie's right leg. After taking off at 8:05 A.M., we linked up with the Interstate highway where it went over the ridge at Rose Lake. From there, it was only 12 miles to the Shoshone County Airport at Kellogg. Landing there would cut the 145-mile leg to Missoula, Montana to 110 miles. The airport was between the Interstate highway and the remains of some mountainside mining operation (it was a superfund site). I landed the Cub with no problem and taxied up to the end hangar that had a large faded painting of a P-51 on its front.

The fellow who managed the airport (a well tanned lanky fellow also named Bernie) came out to meet us. He confirmed that there was no fuel available. Having planned for this contingency, I poured the precious two gallons of fuel into the Cub's tank. Bernie talked to the man for a while and learned that he had based his P-51 here. In that the valley was quite narrow, especially to the southeast, he had to be an excellent pilot to fly a P-51 in and out of this airport. He said that he had sold the airplane to a fellow in Kalispell several years ago.

By now it was after 9:00 A.M. and getting warm. After taking off, I flew to the west side of the rather narrow valley and started a circular climbing pattern. It was necessary to gain some 2,500feet of altitude to get even with the top of the pass. One of our concerns on the return trip was that the west sides of all the passes were rather steep. This meant climbing up to the altitude of the pass, spending considerable time doing that, burning a lot of fuel, and not making much progress along the route of flight. The Cub climbed up to 5,000feet with little difficulty. Once there, we headed up the valley toward Lookout Pass some 25 miles away.

As the Cub went toward the pass, I kept making S turns while climbing to an altitude of 5,200feet. Just over the pass, there was a red and white helicopter below us that was lifting long logs off the mountain side and depositing them next to the highway. Bernie and I had seen this helicopter on the way out, but could not agree on what it was doing. Now we knew.

Once over the pass, the rest of the flight down the valley to Missoula was uneventful. The Cub flew over the Mineral County airport again. From our new perspective, I could see that the trees were at the sides not the ends of the runway. We came out of the valley six miles northwest of the Missoula airport, called in, landed, and taxied to the ramp. It was 12 noon, so Bernie and I walked a few blocks to the airline terminal and had lunch.

When we took off, at 1:35 P.M., I informed the tower that we were going south and wanted a right-hand departure. The tower asked me if I was going over Lola Pass and I didn't have the foggiest idea where the pass was. I told him yes and also informed him that I would be circling for altitude just to the east of the airport.

After a while, the tower called me and asked if I really wanted to circle where I was. I said yes as we were going down the valley to Butte. In a rather disgusted tone of voice, the tower operator said that Butte was to the east, despite it being south of Missoula on the map. He then pointed out where the Interstate went into the hills just to the left of a white M on the side of the mountain. That was the last I heard from him. I'll bet he thought we were a couple of idiots, but you can get away with it in a Cub. Nobody expects a Cub to be far from home. If it is, the pilot usually is completely out of his element.

The truth of the matter was that in the mountain valleys I just could not keep my orientation of what was north-south and east-west to correspond with reality. On the way home, whatever direction I just came from was north. I made orientation mistakes numerous times on the trip. I never seemed to get directions straight until I could see section lines again. On the other hand, pilots who live in the mountains complain about the Midwest as there are no mountain peaks that can be seen 40-50 miles away.

The plan was to fly from Missoula to Bowman Field, refuel there, and then land at Butte for an overnight stay. Bernie and I departed Missoula at 1:35 P.M.. Flying through the valley, our line of flight suddenly appeared to be blocked by a huge set of power lines crossing the valley. We were quite concerned until we realized they were the lines across a side canyon that we had seen on the way up to Missoula. Just as we approached the power lines, the main valley made a sharp left turn and they no longer blocked our path.

About this point, we started having trouble maintaining our height above the ground—density altitude again. The Cub went over a ridge, I found some lift behind it, and was able to climb up to 5,300feet. A few minutes later, all the altitude gained was lost. This posed a real problem as the Bowman Airport was at 5,030feet and Butte was at 5,545feet. In addition, the further south we went, the hotter it was getting. The thermometer in the airplane was reading between 90 and 95 degrees. The Cub was maintaining 4,900feet but not an inch higher. A few miles past Deer Lodge, I realized we would not get high enough to reach Bowman Field. I turned the Cub back towards the Deer Lodge airport, which was somewhat lower at 4,689feet, where we landed at 3:05 P.M..

Bernie and I sat in the 100-degree heat trying to decide what to do. My plan was to stay at the unattended airport until 7:30 P.M. in hopes that it would cool off enough to get to Butte. After a period of time, a woman came to the airport to see if her husband had returned. I learned that she was a friend of the ramrod straight National Guard sergeant who had taken Ann and me into town for gas 20 years ago. She offered to take us into town and indicated that after dinner the Ron Scharff restaurant staff would take us back to the airport. We accepted her offer and she dropped us off at the restaurant. Bernie and I went in and had a couple glasses of lemonade and chatted with Ron Scharff himself. He said that his motel was right next door. If we decided to stay, it would be a good place to RON. This would also put us in a good position to go over both Butte and Bozeman passes in the cool of the morning.

Bernie and I walked around town for a while and visited the old Montana State Prison built in the 1880's. The bank thermometer reading of 101 degrees convinced us that Ron had a very good idea. We checked into the motel and an employee drove us out to the airport to pick up our backpacks. Ron also told us that the county sheriff would take us back to the airport anytime we wanted to go. We had the old lady at the motel desk call the sheriff's office. They said to call again in the morning to remind them of the need for transportation.

Bernie and I arose at 4:00 A.M. and consumed the usual grocery store breakfast items. At 4:50 A.M., I called the sheriff's office and told the woman dispatcher that we needed a ride to the airport. She told me the deputy had just left on a burglary call at a town some 40 miles way and would not be back until around 7 A.M.. In a very disappointed tone of voice, I told her that we really needed to get off early to get over the mountain passes in our antique airplane. She said that she was sorry and hung up. Five minutes later, the phone rang. The dispatcher said that she had called the deputy back and he would pick us up in seven minutes. Sure enough, right on time, he appeared and took us to the airport. From the back seat of the squad car, Bernie told him "it's been a long time since I've ridden back here". The deputy let that one go bye.

The Cub got off the ground at 5:45 A.M., under clear skies, and we flew the 20 miles to Bowman Field at Anaconda. Again, the credit card gas pump was a life saver as we could get fuel at an unattended airport. Nancy Kelley had been surprised that we obtained fuel earlier at Bowman Field. The gas pumps had been out of order for at least a month. No gas would have been a disaster for us as the town of Anaconda was many miles from the airport.

Fortunately, the pumps were working. As we took off from Bowman Field, at 6:30 A.M., the sun was just starting to peek over the mountain tops. The eastern sky was a golden yellow color that faded to light blue high in the sky. The sun's rays were highlighting the tips of the mountain tops while down in the valley it was still semi-dark. The Cub was circled over the airport until 6,300feet was reached. Then, Bernie and I headed toward Butte. From our position, just east of Bowman Field, I could see the saddle where the Interstate highway went over Butte Pass. By the time the Cub was over the Butte airport, the altimeter read 7,000feet. We had plenty of altitude to clear the pass. Looking down at the airport, I could see the morning's commuter airliners taking off. Bernie and I could also see a long strip of green down in the depression that the Cub had flown through earlier.

Once over the pass, the Interstate was followed for about 20 miles. Then, I turned the Cub to the right and flew into the mountain basin that went to Bozeman. We reached there at 7:57 A.M.. This time the Exxon dealer was used as they had a restroom, which reminded me of the three basic rules of aviation:

1.Don't land short.
2.Always fill your tanks.
3.Never walk past a restroom.

The lineboy told me that he was from Madison. We talked for a while about what he was doing in Bozeman.

Bernie and I took off at 8:29 A.M. and then flew the Cub toward the mountainside the glider fellow had mentioned earlier. I circled the Cub over some green fields while climbing up to 6,500feet. Then, the Cub was flown alongside the mountain for a while, but we did not find any lift. Using the engine alone, 7,000feet was reached. Then, I headed the Cub over the saddle to the left of the V. As we approached the saddle, a Piper Vagabond flew past us at what appeared to be only a hundred feet above the terrain. Unlike the Cessna 172 that flew under Ann and me, the Vagabond was off to our left. Once over the pass, we flew past the butte that had been such a help on the way out. Since it had taken so little time to get over Bozeman Pass, Bernie and I elected to by pass Big Timber and go straight to Laurel, Montana.

At Laurel, the Cub's fuel tank was filled. When we tried to start the engine, it would not even pop. The usual starting drills were to no avail. One of the fellows at the airport came over and helped us for a while, but could not make any progress either. This fellow had built two airplanes that consisted of a Super Cub fuselage with Taylorcraft wings and tail feathers. He called them Taylor Cubs. One was silver and the other was red and white. Because of the semi-symmetrical airfoil, the wings had to be installed at a higher angle of incidence than normal. A retired woman school teacher, who was with the fellow, owned a hangar and a red Super Cub. Laurel seemed to be the Super Cub capital of the west. Nearly every hangar had at least one.

I asked the fellow about the large increase in the number of hangars since Bill Vogelsang and I had been here nine years ago. He told us that a new airport manager at Billings decided to get rid of all the general aviation aircraft based there. He had all the city owned general aviation hangars torn down, then pressured the former occupants to leave. As a result, many of the aircraft were moved to Laurel where new hangars were built. The end of the story was that the Billings City Council realized that the manager had made a mistake. Now, they were trying to lure general aviation back to the airport. They built some new hangars, but the hangar rent was several times what it had been in the past. Needless-to-say, they had few takers.

Bernie and I gave up on getting the engine started. We resigned ourselves to waiting for it to cool down. For some unknown reason, we never did pour cold water on the mags. The fellow with the Taylor Cubs asked me about my magnetos. I told him they were shoe box Bendix with no impulse couplers. His only comment was that we had everything working against us.

We had landed at about 10:30 A.M.. Now it was approaching 1:00 P.M.. The day was slipping away from us. I gave the engine another try and lo-and-behold it started. Bernie and I quickly untied the Cub, hopped in and taxied out to the runway. The little thermometer in the Cub was reading in the low 90's, but it didn't feel that hot. The runway was only 3,800feet long at an altitude of 3,815feet. A combination that made me concerned about getting off the ground. The Cub rolled down the runway to the southeast, it lifted off nicely, and started a sluggish climb. As soon as I could, I headed for some green fields and started a circular climb pattern.

When an altitude of 5,000feet was reached, the Cub was pointed at the high ridge just to the south of town and across the Yellowstone River. It was the first of many ridges on the way to Sheridan, Wyoming. The map showed a number of high points of land on or near our line of flight. I tried to use them to help stay on course, but was never quite sure of what constituted a high point. Eventually, Bernie found a highway that was on the map. Given this check point, he was able to determine that we actually were on course. I used a high bump on the horizon as an aiming point, but I never could find it on the map.

At 45 miles out of Laurel, we could see what looked like several arms of canyons to the west of us. I had never flown this leg at such a high altitude. Hence, it was the first time I could see very far into the mountains. As the Cub approached the high bump, we could see the Fort Smith airport and, in a few minutes, the Yellow Tail dam. The canyons we had seen were behind the dam. The bump turned out to be a high point, just to the south side of the dam, that was covered with power lines. Fortunately, two of them paralleled our course line to Sheridan.

After passing the bump, Bernie used the power lines, one on each side of us, to set up our compass heading and proceeded on course. Eventually the Interstate could be seen and as usual there was a final ridge to climb over. In my eagerness to get to Sheridan, I thought each one of several small towns was Sheridan. After several disappointments, the Cub finally reached Sheridan. It was 3:00 P.M. and very hot (100 degrees). Bernie was of the opinion that we should keep on going, but we had been up since 4:00 A.M. and I was very tired. It did not take much to get Bernie to concur that we should stop for the day.

The Holiday Inn was called. In due course, a green Chevrolet Suburban appeared and took Bernie and I to a very large motel. The driver told us that there was a Perkins fast-food place a few blocks away where we could get an early breakfast. After checking in, Bernie and I started to walk to the restaurant that actually was over a mile away. In the intense heat of the late afternoon, it was too far to walk.

Bernie and I awoke at 4:00 A.M. and were at the airport before dawn. This morning, the air felt quite cool. The thermometer in the Cub was reading in the low 70's. There was a high cloud shield oriented from west to east. North of the cloud line, the sky was bright blue and to the south of it there were solid high clouds.

As I taxied the Cub out to the short 23 runway, there was a 10mph crosswind from the northwest. The takeoff was a repeat of the one on the trip out. At about 10 to 15 feet in the air, the Cub quit climbing. I made yet another flat rudder turn to the north and flew along the main runway again. Once past the end of the runway and over the valley, the Cub climbed nicely. In a few minutes, Bernie and I were on the course to Gillette. As temperature was not a factor, the only explanation for the weird takeoff was a downdraft from the wind coming up the face of the cliff and swirling.

A few miles south of Sheridan, Bernie asked if I saw the radio tower. I looked down on the right side. The Cub was only about 50 feet above and 50 feet to the left of the tower. I had never seen the tower and it was not on the map--shades of Dodge Center.

On this leg, Bernie was determined to find the town of Ulm that had eluded us earlier. This time the railroad track was found and we flew along it for quite some distance. Finally, it turned sharply to the left where it followed a large creek to the northeast and then disappeared from view. Bernie never did see Ulm. A few ranch houses near the railroad track was all that was ever sighted. Ulm will remain one of life's little mysteries.

Approaching Gillette from the northwest provided a new perspective on the countryside. I anticipated seeing the city first, but the airport appeared before the city did. Bernie set up the radio frequencies. I called in and we landed at 7:00 A.M.. Much to our surprise it was quite cool. The lineboy said that a front had passed through during the night with some rain and thunder. The edge of the cloud shield remained off to the north.

Bernie and I walked over to the airline terminal and for once had a decent breakfast.

Takeoff was at 7:55 A.M. and the Cub was pointed east in the direction of the Black Hills. This time I was smarter, staying in the valley below the high altitude route taken by Interstate 90. By the time we reached the ridges near Spearfish, the Cub had plenty of altitude and fuel, so we flew to the Sturgis, South Dakota airport located a few miles south of Bear Butte.

The local A&P came out to refuel the Cub and warned me not to use cowl clips behind the prop as they will gouge the back of the prop. He had told Ann and me exactly the same thing when we landed here 20 years earlier. Bernie and I had hoped that by going to Sturgis we could make Pierre, but it still was too far. We had to fly to Philip again.

Earlier in the day I had promised Bernie that as soon as we were out of the mountains, he could be the pilot again. I was very happy to be in the back seat where I could be navigator and stretch my legs (that is stretch in relative terms). It was 91 miles from Sturgis to Philip over open country. There were enough creeks, roads, and power lines to keep us on course. Under the high cloud shield there was excellent visibility. We could see Philip's water tower from some 20 miles out. When we landed, the airport operator was mowing grass, which greatly relieved us. He also was a wheat farmer who could have been combining some distant field. Fortunately, he had finished bringing in the wheat. I decided not to tell him about our problems with his dirty gas as it was a bad call on our part.

I went back in the front seat to be the pilot on the short 69 mile leg to Pierre. The Cub broke ground at 12:39 P.M. Approaching the city, there should have been a power line coming in from the right, however, I could not see it. At 10 miles out, Highway 14 was seen and Bernie determined that the Cub was about 4 miles north of course.

After landing and refueling the Cub, I asked the lineboy where we could get some lunch. He offered us the use of their courtesy car to go to a shopping mall three miles down the road toward town. Bernie and I each had a sub sandwich that was just better than awful.

At Pierre, Bernie went back in the front seat. The next stop was Huron--99 miles away. Once off the ground and at our cruising altitude of 300 feet. I realized the course line had a heading of exactly due east. I told Bernie he could follow the east-west road below us as far as it went. When the road ended, Bernie moved the Cub over a few hundred yards and followed fence lines. When they ran out, he found another road. After an hour and 37 minutes we landed at Huron at 5:30 P.M.. By now both of us were very tired, so we decided to RON here.

The lineboy was a young fellow, named Lance, from La Crosse, Wisconsin. He had graduated from the flight training program at the Mankato, Minnesota State University. As he was working toward an airline career, he came to Huron on the promise of getting considerable multi-engine time. The flight hours had not materialized, so today was his last day on the job. He was moving to Watertown, South Dakota for a position that offered more flight time. Lance called a local motel and reserved a room for us at the corporate rate. The motel promised to come out and pick us up.

While we were waiting, a charter flight came in for fuel. Four guys and a girl came into the office. At first I though they were a rock band. Finally, I noticed they had on T-shirts that said Excelsior-Henderson motorcycles. I pointed this out to Bernie who then struck up a conversation with them about their cruising bikes. Several weeks later, I saw their picture in a Forbes magazine article about their company. The fellows we talked to were the principles in the company. A year later, I read that the company had gone bankrupt and was no more.

Bernie and I waited, but the motel van never appeared. Finally, when Lance closed the office at 7:00 P.M., he drove us to the motel. I was very impressed with this young man. I knew that some day he will make an airline a valued employee. He just as easily could have been a disgruntled employee and given us the minimal amount of service, instead, he treated us royally.

Leaving Huron the next morning(Wednesday), I pulled a dumb trick. As I checked the mags, my finger became entangled in the key ring and I turned off the engine! I quickly jumped out, flipped the prop, and much to my delight it started on the first flip. The trip to Marshall, Minnesota was uneventful as we were flying low over farm land with section lines every mile.

Near the Minnesota border, I thought I could see a tall power line to the right of course. Upon closer inspection, it became apparent that the power line actually was a row of twenty wind-driven electrical generators. Initially, I thought they had strobe lights on them, but it turned out to be the sun reflecting off the propellor blades. Bernie told me the generators were perched on Buffalo Ridge that acts as a dividing line between rivers flowing north to Hudson's Bay or south to the Gulf of Mexico.

As the Cub turned base leg to land at Marshall, a twin engine Piper Seneca, on final approach, went right in front of us. I had not seen it before then. He might have called in, but our radio was not turned on (you are never alone). He landed long and I landed short while he was still on the runway. At the gas pumps, we talked to the young fellow who was the pilot of the Seneca. He said that when he had turned around on the runway to taxi back, he saw a plane on final approach. Recognizing that it was a Cub, he just waited until it landed and was clear of the runway. The Seneca was equipped with a digital camera system for photographing farmland. He also said that he had flown a Cub quite a bit as his dad owns one. Checking our logs, it took us an hour and 58 minutes to fly the 117 miles for a ground speed of about 60 MPH, which is not bad.

The next leg was to Mankato, 94 miles away, with Bernie in the front seat. From the newspaper reports, we knew that New Ulm had suffered considerable tree damage from a recent thunderstorm, but from the air we could not detect any. Bernie deviated from our course line here in order to fly to St. Peter to look at the tornado damage that had occurred earlier in the summer. He had seen it from the ground and now wanted to see it from the air. Some buildings had new roofs, many basements had no houses on them, and the Gustavus Adolphus College campus had no trees. Many campus buildings had their windows blown out. The adjacent farm fields were still strewn with lumber, insulation, and pieces of sheet metal roofing.

After crossing the Minnesota River, Bernie turned the Cub southwest toward the Mankato airport only a few miles away. As the Cub entered the downwind leg of the traffic pattern, we heard someone call in as flight 3521 turning base. Another one called in as flight 3544 on final. We started looking around for commuter airliners, but all we saw in the traffic pattern were Cessna 172's. It turned out that the Civil Air Patrol was having a solo camp here. The student pilots were calling CAP flight numbers.

The path from Mankato to La Crosse, Wisconsin was one that I had flown many times. Due to our 72 Mph ground speed the checkpoints along the 136 mile course just seemed to flow under us. As usual, we flew over Ed Schermerhorn's house, but did not see anyone outside.

Approaching the Mississippi River bluffs, I called the La Crosse tower on the old radio frequency. After some complaining by the female tower operator, she gave us a right-hand approach to runway 22. The 15mph wind was nearly perpendicular to the runway, so I decided to make a wheel landing. Just as the Cub's left wheel touched down, the wind swung around to our right rear quarter. I bounced the landing, not once but three times. Virtually, a repeat of my horrible landing at The Dalles. It took me awhile to get the Cub calmed down enough to stay on the ground.

As I was taxing along the ramp, the tower operator yelled at me:

"Don't turn there."

I replied:"If I don't S turn, I can't see where I am going."

Apparently, she had never seen a Cub taxi.

I went into the pilot's lounge to check the weather. There were two Air Force pilots(a Major and a Captain) trying to get the Weathermation computer system to work. I showed them how to do a few simple things (all that I knew). They were flying a Lear Jet and had been from Colorado Springs to Los Angles and then to La Crosse in less time than it had taken Bernie and I to fly from Huron to La Crosse. The Major asked me what I was flying. I told him that we were returning from Seattle in a Piper Cub. He asked me what else I flew. I said that the J-3 was all that I flew. He just shook his head and made no further comment, but it was clear that he thought I was not playing with a full deck.

On the radar graphics, there was a line of rain storms across our path to Waunakee. The red and yellow areas on the computer screen indicated a narrow band of heavy rain. It was Bernie's turn to fly the front seat, but I decided I'd rather be there if we encountered bad weather. We took off to the east and entered the La Crosse River valley and then climbed up to the convent at St. Joseph. From there, everywhere to the east looked dark, but visibility was good and the ceiling was decent.

About 2/3 of the way to Waunakee, we flew into some light rain that quickly became heavy. I could see a light spot on the edge of the rain shower, so I headed for it. Once out of the shower, the Cub was flown only a hundred yards from the edge of the heavy rain that looked like a waterfall coming out of the dark gray cloud base. Once around the rain shower, I mistook a hump on the horizon for the Baraboo Bluffs. I pointed the Cub's nose at the hump. Ten minutes later, we flew over a town that did not look like any on our course to Waunakee. Suddenly, I recognized the large Catholic Church, high on a hill, in the town of Plain. The Cub was 10 miles off the course--so much for the 100-yard rule. I corrected the course and after a flight of one hour and 23 minutes Bernie and I landed at Waunakee.

As the dark line of clouds could be seen just north of the Wisconsin River, I expected the rain to hit Waunakee in about ten minutes. By the time the Cub was tied down in the hangar, the line of rain showers had dissipated.

While Bernie and I were on our trip, JoAnn had flown the airlines to San Jose, California to visit our daughter, Ann. Since we were home earlier than expected, she was not there to pick us up. I had noticed Sally Meier working in her yard and asked her if I could use her phone. I called a couple of people to get a ride home, but nobody answered their phone. Sally graciously offered to take us home and we were there by 5:30 P.M.

Postscript

Bernie and I had been gone 16 days, flown 3,969 miles, that is the same distance as New York to Berlin, Germany. Sixty-five hours of flying time had been accumulated at an average ground speed of 61MPH. While not a national record for long Cub trips, it was the furthest that I had ever flown. The Cub had been over a wide variety of terrain: Midwestern farmland, mile after mile of prairie, seemingly endless ridges, down long mountain valleys, over mountain passes, across desolate landscapes, along rivers, and tree-covered mountains. While many other airplanes can do this with ease, it is somehow different in a Piper Cub. There is an intimacy with the land that alters one's perception of flying. Most of the time is spent flying only a few hundred feet off the ground where one can observe the features of the landscape in great detail. In good weather, the visibility is such that you can often see 50 or more miles and it seems to take forever to get to the next checkpoint.

The main feature of the trip to Seattle was the unrelenting heat. West of Rapid City, nearly every city was reporting very high afternoon temperatures, many were near or over the 100-degree mark. As a result, trying to gain altitude was the major effort of every day. Some times, it was acquired easily, but other times it was an endless battle to gain and maintain a few dozen feet. The takeoff at Butte was the second closest I have ever come to crashing the Cub. I was within a millisecond or so of telling Bernie to prepare himself for a crash, but the old admonition "fly the airplane first" won out.

The only redeeming feature of the incessant heat was that the air was smooth much of the time. Even in the relatively cooler weather, after Gillette on the way home, the winds were light. This was a great relief as I have often flown in 30mph or more winds in the plains states. Such winds can make a trip a lot less enjoyable.

Before the trip, I was quite concerned about getting over the mountain passes. On the way to Seattle, the Cub cleared Bozeman, Butte, and Lookout passes by only a hundred feet or so each. Had there been even moderate winds, we would have had a much more difficult time getting over them. On the way home getting up early in the morning allowed us to fly over the passes in the cool of the day when gaining altitude was not a serious problem. The Cub cleared the three passes with plenty of room.

At the beginning of the trip, Bernie mentioned that he had made out a new will. At the end of the trip, he said that he felt a little sad that he had not been able to use it. Had circumstances been slightly different on this trip, he would have had several opportunities to do so.

Yellowtail Dam

Columbia River South of Ephrata, Washington

Snowqualmie Pass in Washington

Sunrise at Sheridan, Wyoming

Epilogue

During the summer months, many Saturday mornings are spent working on the Cub. Sometimes making minor mechanical repairs, but often just washing it. Inevitable, some fellow in his 60's or older will stop by to tell me that he learned to fly in a J-3. There is always an underlying tone of nostalgia in their voices as they tell about flying the Cub. Some of these men subsequently had flown for many years. Others may not have flown since their youth. Yet, there is a common bond of having shared an experience that only other Cub pilots can truly relate to. They do not talk about the Cub being underpowered, cruising at a sedate speed, how uncomfortable the seats were, or that looking forward all they could see was the back of the instructor's neck. What they do reminisce about is how they learned to fly a J-3. While the Cub is quite mild mannered in the air, it is quite demanding when one is landing or taking off. A moment's inattention to directional control, can result in a ground loop or sudden trip through the grass along side the runway. I suspect that in order to learn to fly in a Cub, one must develop an emotional bond with the airplane before it can truly be mastered. This bond, which is a part of the Piper Cub mystique, seems to stay with the pilots for the rest of their lives. I have talked to many people who have learned to fly in Aeroncas , Taylorcrafts and Cessnas of the same era as the Cub, but they all seem to lack an emotional bond to these aircraft. In my own case, I first flew a Cub when I was 17 years old and the experience was well integrated into my psyche. When the Cub was restored in 1969, I was able to fly it solo after only a minimum of instruction. It seemed as if I had flown a Cub just a few months ago even though it actually had been some 22 years earlier.

Over a period of time, the Cub became a well trusted friend. Each time I climb into the cockpit, it is like joining an old pal for an afternoon. Instinctively, I scan the few instruments, check the positions of the switches, and wiggle a bit on the seat cushion until I feel at home. By now I know most of the Cub's idiosyncrasies both mechanical and aerodynamic and have developed an inner sense of what is normal. During flight a small vibration would be detected that felt different. A post flight inspection would reveal that the wooden propeller had dried out and needed to have its bolts tightened. The tail wheel would shimmy on a hard surface runway because the tail wheel post bushing had worn.

The instruments in the Cub were old when I got them and I have never had them overhauled. The tachometer always reads 120 RPM too low, so I cruise at 2,000RPM rather than 2,150 on the dial. The altimeter reads properly, but the barometric pressure readings in the small window have no relationship to the current barometer settings. The airspeed instrument seems accurate, but is not really needed as I have developed a sense of how fast the Cub is going over the ground. The crucial oil pressure and temperature gauge has always worked perfectly, but the engine primer has never done it's job effectively. A purist would probably have a fit if they looked at the Cub closely, but these are small things that have no effect upon how the Cub flies nor on its performance. Yet they are part of what gives the plane its personality. A truism among Cub owners it that no two Cubs are alike. It is details like these that make the truism believable.

When I restored the Cub in 1967-69, it was covered with Ceconite fabric that is very durable and can last twenty five years or more. While this is a desirable feature, it also can hide corrosion of the underlying steel fuselage framework. After flying the Cub for twenty one years, I became concerned about the integrity of the fuselage tubing as rust was showing on the tops of some lower tubes. The second week in September, 1991, my crew of friends and I disassembled the Cub and hauled it home on a trailer. With great reluctance, I took a razor blade to the fabric and removed the covering. Two squares of fabric with the Cub's bear decals on them were cut out and saved. JoAnn had both of them framed. One was given to Ann as a memento of our adventures. The other hangs on my bedroom wall. The paint was removed from the fuselage framework and in the process I discovered a hole in the lower right longeron. The hole was at the bottom of a tubing cluster three feet behind the wing and a 6 inch section of longeron was replaced. Finding that hole justified tearing the airplane down. One change I did make was to paint the aircraft's N number in 3" letters on the rudder rather than the easily seen 12" letters on the side of the fuselage. By the middle of May the second restoration was finished. The same crew took the Cub back out to the Waunakee airport and reassembled it. My A& I re-certified the Cub and it was ready to fly again.

Starting the Cub's Continental A65-8 engine has always been an art form rather than a science. For many years, the engine did not have impulse magnetos that aid in starting the engine. When the engine is cold, pulling it through 8-12 propeller blades will suck gas into the cylinders. Once the magneto switch is turned on, one quick flip of the prop will start the engine. There is a trick to this flip that insures the engine will start on the first blade. Whenever, a stranger tries to start the engine it is a sure thing that it will not start. Often, they can flip it for 10 minutes or more to no avail. I will reestablish the starting conditions and it will start first flip for me (It's all in the wrist). There was also a ritual for starting the engine when it is hot. Pull the engine through exactly two blades, set the throttle about 3/8" forward, turn the magneto switch on and use the magic flip. There are days when the engine does not start on this first flip. Then it is to the backup drill. Turn off the magneto switch, throttle full forward, pull the prop through 16 blades, pull the throttle back, advance it 3/8", turn the switch on, and give a propellor blade a quick flip. Nine times out of ten, the engine will spring to life whereupon I reach into the cockpit and retard the throttle. It is imperative to mention the all of this is done from behind the propeller with my left foot jammed under the right wheel and my left hand firmly grasping the front of the door opening. Despite all this, I have started the engine at full throttle three times. The minute I hear the full throated roar, I know that I have made a serious mistake. A mad dive into the cockpit ensues where I slap the throttle back as the Cub starts to roll over my left foot. In all three cases, I was distracted by someone talking to me in the final stages of the backup engine starting drill. After the 1998 Seattle trip, I could not tolerate the starting problems any longer. A new pair of Slick magnetos, with impulse couplers, were installed. Now, the engine starts first flip hot or cold and the secret flip of the wrist is not longer needed. I should have done this ten years ago.

The continuing saga of the radio has been a sore point for many years. Many hundreds of dollars and untold hours have been spent trying to get a radio to work well. The 10 crystal Regency handheld radio, mounted in a bracket on the back of the front seat, always was marginal and difficult to use. The Regency radio was replaced by a NARCO Com 10A radio. In the beginning it worked flawlessly, but after ten years of use, it was marginal. It was replaced by a Mitchell TKM 11 radio that initially gave me some problems. It was sent to the factory twice before it worked without sounding like someone was sending Morse code. The kit built intercom was replaced by a Sigtronics 400A intercom. This unit was designed for a high noise environment, such as the Cub's cockpit. At this time, I also installed all new wiring and three sets of interconnection plugs mounted in a bracket under the front seat. Now the whole radio system can be removed by disconnecting three plugs. Since then, the radio has worked perfectly.

When flying locally, I do not have the radio installed in the Cub as it seems sacrilegious. When I am going on a cross-country trip, I slide the radio into it's tray to cope with the occasional control tower and sometimes to confer with an accompanying airplane. I have gone flying with a friend of mine who has a Mooney 231. From the minute he climbs into the cockpit until he climbs back out he is talking to someone on the radio. There are no more than 30 seconds when he is not talking to ATC, the tower, or anybody that will respond. He drives me nuts. In the Cub, the only sound that I really want to hear is the steady drone of the Continental A-65. To me going up in the sky in the Cub is like entering a vast Cathedral where I am the only parishioner. What I want to do is savor the distant horizon, admire the colors, smell the farm fields, and feel the currents of air. What I do not want to do is listen to the incessant inane chatter that goes on over the common radio frequencies. I like it even less when a control tower is telling me what to do and I only can understand every third word. It seems they have a special school where they teach control tower operators to talk as fast as they can. Usually, I pretend to understand and just keep doing what I think is the normal thing to do.

In the days when the PATCO union existed, many of the control tower operators were downright nasty. Once I landed at Eppley Field, Omaha, Nebraska. The tower operator harassed me unmercifully for having the audacity to land a Cub at an airport served by the airlines. After President Reagan fired the PATCO members, the tower operators became much more civil. Most of the larger airports now have a MODE C requirement, so the Cub and I can not land at them. I do not consider this a major loss as there are still enough smaller airports to accommodate those of us with a minimum of electronics. However, the old time grassroots airports are rapidly disappearing as they become economically unfeasible.

When going on a long cross-country flight, forecasting the weather always is a critical issue. Since these trips involve several days of flying, it is important to understand the trend in the weather. In the early days, the flight service stations were staffed by men who had spent their whole careers taking actual weather observations and learning the idiosyncrasies of the weather. Talking to them usually was very helpful and I was confident in what they told me. In recent years, the briefer's are simply readers of computer printouts. In particular, the briefer's assume that you are going to complete your trip in a few hours and have no interest in the day after tomorrow's weather. One gets the impression that the only weather they know is obtained by laying on the ground and looking straight up. In college, I took a course in meteorology taught by the chief meteorologist of the Manhattan project. For the final exam, we were given the teletype tapes for three days. We plotted the national weather charts ourselves and then had to predict the weather for five cities in various parts of the country. Our grades depended upon the accuracy of our predictions. Because of this training and access to NOAA's weather graphics on my personal computer, I now rely on my own analysis of the weather. I only call the FSS if I have a need for a specific piece of information, at a given location, for the current day.

Having been an Air Force Navigator, before the days of all the electronic gadgets, my first love is cross-country flying. As the Cub and I proceed along the course, I record the time at each checkpoint on the map, compute the ETA to the next checkpoint, and watch the countryside go by. In the early days, I used to compute the wind direction and velocity about every third check point, but have not done that for many years. After leaving the airport, a good initial check point lets me establish the wind correction angle necessary to stay on course. If the wind holds, I simply fly along the course line at that angle. Usually this angle is a modest 10-15 degrees, but I have some movies taken while flying along an interstate highway in Nebraska with a 45 degree wind correction angle and a 30 mph ground speed. Occasionally, the wind does not hold steady and things get interesting. Once, while flying east across northern Nebraska, the wind switched three times in less than 100 miles. It can be tough to catch such wind shifts as I don't know if I am just not doing a good job of holding a heading or there really is a shift. If the visibility is good, it is not a real problem, but if the visibility is limited, the Cub can get off course rather easily.

One of the oddities of cross-country flying is that other airplanes are rarely seen. At any given moment in time, there are very few airplanes in what is a huge air space. Most general aviation airplanes are flown at

2,500feet or higher. From the Cub's low altitude they are hard to see unless they are only a few miles away. When I do see another airplane, it is usually in the neighborhood of an airport. As mentioned in the earlier chapters, I have encountered other aircraft at the Cub's cruising altitude. Usually the ubiquitous Cessna 172. Despite the rarity of sighting another airplane, I always assume that there are other aircraft nearby. Thus, the "you are never alone" mantra.

I am often asked if it gets boring while flying the Cub on long cross-country flights. Usually, by a pilot who flies his plane at 8,500 feet, on autopilot, and listens to Rock and Roll on his stereo system as he drills holes in the sky under the control of ATC. This question often surprises me as boredom is never an issue for me. Since the Cub is an airplane that must be hand flown at all times the stick and rudder pedals require constant attention. Navigation also consumes considerable time and effort. I have to constantly be scanning the countryside looking for features that are on my maps and verifying checkpoints when they are sighted. Then it is necessary to write the time on the map, measure the distance from the previous checkpoint, get out the E6B, compute the ground speed, and the ETA to the next check point. I also compute the flying time to the final destination and make a quick determination as to whether or not I have enough fuel to get to there. All-in-all, the Cub's cockpit is a very busy place and boredom just doesn't happen. Getting tired, yes, but bored, no.

Radio towers and power lines have always been a concern of mine. The former since the close call on my first solo cross country and the latter from the wires across the Wisconsin River. I have often told my co-pilot/ navigator Bernie that radio towers and power lines were "Cub Grabbers" and after a few stints in the front seat he became a believer. The following quote is from a letter he sent me a few weeks after the 1998 Seattle trip:

> "I could never understand your concern about power lines and radio towers. After all, weren't we always well above them? But, when it came my turn to fly the airplane, I could almost feel the wires reaching up into the sky to snag us. Mark Twain described deadhead in the Mississippi River as "Fishing for Steamboats". I felt like the wires and radio towers were "Fishing for Cubs"

To achieve the illusion that the Cub is flying faster than it really is, I fly low. At a higher altitude, say 1,000feet AGL, the Cub seems suspended in space. You feel as if you are not making any progress at all, which very well may be the case. Down low, the trees, farms and roads seem to come at you from the horizon and slide underneath at a reasonable clip. The ground speed achieved by the Cub is an object of considerable interest to me. Examination of my flight logs shows that the usual ground speed ranges from about 50mph to 65mph. My lifetime average is in the neighborhood of 55mph and that is what I use when planning trips. This gives some allowance for headwinds and normally leaves a small reserve of fuel in the Cub's gas tank. A ground speed of only 12mph has occurred twice, but did not hold for long periods of time. On a few occasions, ground speeds in excess of 72mph were achieved, making some long legs seem short. The highest ground speed ever achieved was 112mph going from Mankato to LaCrosse. Unfortunately, this happened only once.

My Cub does not have a cockpit heater, so I do not fly it in the winter. The wooden propeller is removed on Thanksgiving weekend and reinstalled when the frost is out of the ground in mid-April. One consequence of this practice is that my flying season corresponds to the growing season. In Wisconsin, I get to watch the countryside go from the tired brown look of late spring to the lush green of summer. Later, the green fields of summer turn to the tans of mature corn and soybean. Finally, the woods turn to brilliant yellows and reds interspersed with the dark green of the pine trees. When the leaves fall, the countryside returns to its dull earth tones of early winter and the season is over. Observing this cycle from the Cub gives me a sense of connection with the earth somewhat like that of those who live on the land.

To me the most enjoyable country to fly over is the prairie in Nebraska, South Dakota, and Colorado. The land flows in a gently undulating manner towards the horizon. The high points are well rounded and the valleys between the higher land are also gently dished out. As you fly along, the Cub gets in rhythm with the land, rising up as it lifts and dropping down as it lowers. Even though the visibility may be 30 miles or more, there is nothing on the horizon that can be used as a goal. Navigation depends upon the shape of a ridge, an occasional creek, a railroad track, a dirt road, and the rare village. The distance at which you can identify landmarks is limited to only a few miles, so I have to pay careful attention to the nearby countryside as it flows by. It is easy to miss a checkpoint that is only a half mile away. Sometimes it is behind a hill, other times it is in a small valley. Often I am looking out the left side when it slipped by on the right. Despite the vastness of this country I feel quite at home knowing that if I fly long enough in the right direction the next airport will appear at the proper time and place. The position of the fuel gauge wire is a good indicator of how long the Cub and I have been flying. If the bent wire looks too low for the amount of distance yet to go, a certain level of concern arises.

Flying in the mountainous western states is a totally different experience. The skies are usually clear, the visibility is more than fifty miles, and mountain peaks are the reference points. A peak on each side of the course is sufficient to establish the basic line of flight. Map reading the ground details such as edges of mesas, a few dry creeks, and the occasional road are all that is necessary to fine tune the compass heading and keep on the course line. Many times, the Cub is down in a mountain valley and I am looking up at the sides of the mountains. Navigation then consists of simply following the highway, at the bottom of the valley, to my destination.

It should be mentioned, that there have been many a cross-country flight where the Cub and I had totally different opinions as to where we were going. I would set up a compass heading and proceed along the course line. In ten minutes or so, the Cub would be pointing 10 to 15-degrees off the desired heading. The heading would be corrected and watched for a few minutes to see that the Cub was maintaining it. Then my attention would be turned to other things. In another ten minutes or so, I would become aware that the Cub again was off the heading by the same 10 to15-degrees. A whole day can be spent playing this game with the Cub. I have never discovered the reason for this Cub behavior. I also have never been able to predict when it is going to happen.

When flying low, you can easily observe people on the ground. Many people do not even hear the Cub as it goes by. Small children and farm wives generally will wave to me as I fly overhead. Amish children in particular get quite excited and will wave at me for quite a while. When I notice them, I always wave my arm and give the wings a good waggle. From the expressions on their faces they seem to enjoy the response from the little airplane. On one trip to Denver, a farmer was putting new shingles on the roof of his house. As I flew over him, he gave me a friendly wave, hammer in hand. Returning a week later I flew over the same house. He was still at work and he waved to me again. Once Bernie and I were flying the Cub across Indiana. When we flew low past a barn. Two large dogs came running out of the barn and chased the Cub across a field. If they had caught up with us, it would have been very embarrassing.

In the ranch country of Nebraska, South Dakota, and Colorado, the cattle are interesting to watch. As the Cub flies past, the old cattle just turn their heads and track the plane for a few seconds before returning to their grazing. The young cattle will always scatter and run a few hundred feet before stopping. Calf's usually run towards their mothers and try to hide.

Birds are another story altogether, Hawks, Eagles, Sea Gulls, Geese, and Buzzards think they own the sky. The Cub will approach one of these birds and they will not move an inch. Many is the time that a Hawk will simply turn it's head and watch me as I pass only a few yards from it's face. They seem completely unperturbed by this big machine. In the spring and fall, Sea Gulls migrate through Wisconsin. Huge clouds of them will be riding a thermal or circling to land. They are extremely dangerous as they are: big, fly quite high, there are lots of them, and you can't keep track of all of them. I typically try to give Gulls a very wide berth and will even fly around for a while until they decide to move somewhere else. Once I was landing, on a cold, windy day at Albert Lea, Minnesota. A flock of gulls had decided that the sun warmed asphalt runway was a good spot to rest. I made a low pass to scare them off, but by the time I got around on final again they were all back on the runway. I landed short and as I taxied up to them they lifted off.

One area where the Cub has a significant advantage over other airplanes is in the area of flying in poor weather conditions. When the ceiling gets below a thousand feet and the visibility is less than 3 miles, this is great Cub weather. In these conditions, I feel as if I am at the center of a grey cocoon that seems to move along with me. Typically, the clouds have blocked out the sun so that within this cocoon the sky is a medium to dark grey. In many cases, there may also be heavy haze that adds to the feeling of being out of contact with the world. Although I can not see very far in any given direction, the visibility straight down and for a short distance forward is usually reasonable. The Cub is slow enough that I have enough time to study the ground as it passes by so that I can still navigate. Sometimes, the Cub and I have flown for the better part of a day with the cocoon seemingly inching its way through very smooth air and we stayed right on course. It was simply a good day to fly in which the visibility was limited. These are days that I truly enjoy as I know most other pilots are sitting on the ground cursing the weather while I am merrily flying on my way.

On other days, navigating under these conditions becomes a major task. When the visibility gets down to less than a mile and the ceiling is only a few hundred feet the cocoon shrinks to an uncomfortably small sphere. I spend most of the time peering forward into the gloom to try to see if there are any tall trees, radio towers, hills, etc that are going to suddenly appear out of the grey. My experience has been that the compass seems to start to swing about 45 degrees side to side. I think that the Cub is flying a constant heading and the compass is acting strange, however, just the opposite is the case. I am turning this way and that way to try to see ahead, which results in not holding the heading. In addition, these excursions are not of equal amplitude and after a few minutes, the Cub is off course to an unknown degree. The navigational difficulties are compounded by the amount of time a check point can be seen as they are visible in the gloom for only a few seconds. It is not feasible to fly back to recheck a landmark as I may or may not find it again. The nice thing about a Cub is that when things get too sticky, I can put it down on the nearest level spot.

I decided many years ago to never land on a highway. Even if there are no wires alongside or across it, there are always road signs, mail boxes, etc to snag the unwary. In the early days of aviation, landing in a farmer's field was often a dangerous affair due to hidden rocks and holes and many an expensive propeller was broken. Due to the cost of modern farm machinery, the typical farmer keeps his fields in a reasonably flat condition and free of big rocks although wild animals can dig some good sized holes. When selecting a farm field for a landing, one must pay attention to the color of the field. The richer the green color, the deeper the grass or alfalfa, and the more dangerous the landing. By ignoring this dictum many an unwary pilot has ended up with their airplane on its back in a pretty green field. Over the years, the Cub and I have made use of the "friendly farmer airport" on a five occasions. Three of these landings were due to weather conditions and two were due to mechanical problems. In every case, the farmer has been very gracious about our predicament and helped us in every way possible.

Beginning in the mid-1990's, flying low in bad weather (scud-running) started to become a very dangerous proposition. Cellular telephones were coming into wide spread use. To make them function over wide area, large numbers of towers were erected, especially near urban areas. They also seem to appear at random locations in rural areas. These towers range in height from about 100 feet to much higher. Most, but not all, of the taller towers have a small rather weak strobe light on their tops, but no other markings. The towers are constructed of zinc coated steel that gives them a grey appearance. In low visibility and ceilings, this grey color blends into the background, and the towers become nearly invisible. Often they can not be seen until they are only a few hundred yards away. In addition, they are not depicted on the aviation sectional maps. The net result is that my minimum altitude in bad weather has been raised several hundred feet.

One of the joys of having done so much cross-country flying in the Cub is reminiscing about past flights. Over the years, I have amassed a considerable collection of photographs taken on these flights. A large wall map of the United States, with all the routes taken represented by red lines on it, hangs on a basement wall. These artifacts provide me with reminders of where I have been, who was with me, and what adventures were associated with the trip. The Weather Channel also triggers my memory. When they say that Snoqualmie Pass received a foot of snow, I can visualize the ski area on the ridge and the mountain wall at the end of the ridge. When high winds are reported in eastern Colorado, I can mentally fly the landing at Julesberg all the way to the hangar. Automobile advertisements on TV often show Monument Valley. Seeing the road that Ann and I followed through the spires, causes me to hum a few lines of "Follow the Yellow Brick Road." A large photograph of the Cub, behind the haystack in Montana, hangs on my bedroom wall. Looking at it brings a visualization of flying down the wash gully. Hearing Bernie's voice on the telephone reminds me of looking up at the runway at Butte, Montana.

I find it very difficult to put into words the essence of 37 years and 3,000 hours of "Cubbing." I hope that the body of this manuscript will in some sense reach toward that elusive goal. In reality, the Piper J-3 "Cub" is a rather frail aircraft, it is very light and if not handled properly can be damaged rather easily. At the same time it is also a tough bird ready to take on all kinds of weather and flying conditions. Both on the ground and in the air the silhouette of a Cub is readily identifiable as it has an unique appearance. If nothing else the sedate pace at which it passes by marks it as a Cub. Perhaps the best analogy that I can make is to a beautiful woman of slight build, delicate features and a sparkling personality, but with a strong inner sense of who she is. You must relate to her with care, respect, and a certain amount of awe while enjoying the relationship to the fullest. I think this was best summed up by a fellow who was trying to talk me into selling my Cub to him, so he could turn it over and make a quick dollar. After about a half hour of telling him that I was not ever going to sell my Cub and especially not to him, he grumbled " You Cub owners are all the same--you are in love with your airplane."

He was absolutely right!

References

Barrett, G. (2008). *Is a second referendum appropriate in order to allow Ireland to ratify the Treaty of Lisbon?* Available at SSRN 1263300.

Blais, A., Gidengil, E., & Nevitte, N. (1996). *The challenge of direct democracy: The 1992 Canadian referendum*. Montreal, QC: McGill-Queen's University Press.

Bonde, J.-P. (2010, February). *From EU Constitution to Lisbon Treaty*. Foundation for EU Democracy and the EU Democrats in cooperation with Group for Independence and Democracy in the European Parliament.

Cussen, M., & Phelan, G. (2010). Irish households: Assessing the impact of the economic crisis. *Central Bank Quarterly Bulletin, 4*.

de Brca, G. (2009). If at first you don't succeed: Vote, vote again: Analyzing the second referendum phenomenon in EU treaty change. *Fordham International Law Journal, 33*, 1472.

Dinan, D. (2009). Institutions and governance: Saving the Lisbon Treaty? An Irish solution to a European problem. *JCMS: Journal of Common Market Studies, 47*(s1), 113–132.

Dinan, D. (2011). Governance and institutions: Implementing the Lisbon Treaty in the shadow of the Euro crisis. *JCMS: Journal of Common Market Studies, 49*(s1), 103–121.

Gallego, M., & Schofield, N. (2013). The convergence coefficient across political systems. *The Scientific World Journal*.

Irish National Election Study 2002–2007. http://www.ucd.ie/issda/data/irishnationalelectionstudy/.

Laursen, F. (Ed.). (2012). *The EU's Lisbon Treaty: Institutional choices and implementation*. Surrey, UK: Ashgate Publishing, Ltd.

Millward Brown and Landsdowne (2009a). *First Lisbon Treaty*. Irish Social Sciences Data Archive. http://www.ucd.ie/issda/data/eureferendums.

Millward Brown and Landsdowne (2009b). *Second Lisbon Treaty*. Irish Social Sciences Data Archive. http://www.ucd.ie/issda/data/eureferendums.

O'Brennan, J. (2009). *Ireland says no (again): The 12 June 2008 referendum on the Lisbon Treaty*. Parliamentary Affairs.

O'Malley, E. (2007). The Irish parliamentary elections of 2007: New gloss on an old engine. *West European Politics, 30*(5), 1158–1165.

Quinlan, S. (2009). The Lisbon Treaty referendum 2008. *Irish Political Studies, 24*(1), 107–121.

Schofield, N. (2007). The mean voter theorem: Necessary and sufficient conditions for convergent equilibrium. *Review of Economic Studies, 74*(3), 965–980.

Schofield, N., & Gallego, M. (2011). *Leadership or chaos: The heart and soul of politics*. Springer Science & Business Media.

Schofield, N., & Sened, I. (2006). *Multiparty democracy: Elections and legislative politics*. New York: Cambridge University Press.

Sinnott, R., & Elkink, J. A. (2010). *Attitudes and behaviour in the second referendum on the Treaty of Lisbon*. Report for the Irish Department of Foreign Affairs, University College Dublin.

Stokes, D. E. (1963). Spatial models of party competition. *American Political Science Review, 57*(2), 368–377.

Tonra, B. (2009). The 2009 Irish referendum on the Lisbon Treaty. *Journal of Contemporary European Research, 5*(3), 472–479.

Train, K. (2003). *Discrete choice methods for simulation*. Cambridge, UK: Cambridge University Press.

The manufacturer's authorised representative in the EU is Springer Nature Customer Service Centre GmbH, Europaplatz 3, 69115 Heidelberg, Germany. If you have any concerns regarding our products, please contact ProductSafety@springernature.com

Printed and bound by CPI Group (UK) Ltd, Croydon, CR0 4YY
30/09/2026
02225646-0012

Made in United States
North Haven, CT
05 November 2026

21760242R00115